INVERSE AND ILL-POSED PROBLEMS SERIES

Forward and Inverse Problems for Hyperbolic, Elliptic, and Mixed Type Equations

Also available in the Inverse and Ill-Posed Problems Series:

Nonclassical Linear Volterra Equations of the First Kind
A.S. Apartsyn
Poorly Visible Media in X-ray Tomography
D.S. Anikonov, V.G. Nazarov, and I.V. Prokhorov
Dynamical Inverse Problems of Distributed Systems
V.I. Maksimov
Theory of Linear Ill-Posed Problems and its Applications
V.K. Ivanov, V.V. Vasin and V.P. Tanana
Ill-Posed Internal Boundary Value Problems for the Biharmonic Equation
M.A. Atakhodzhaev
Investigation Methods for Inverse Problems
V.G. Romanov
Operator Theory. Nonclassical Problems
S.G. Pyatkov
Inverse Problems for Partial Differential Equations
Yu.Ya. Belov
Method of Spectral Mappings in the Inverse Problem Theory
V. Yurko
Theory of Linear Optimization
I.I. Eremin
Integral Geometry and Inverse Problems for Kinetic Equations
A.Kh. Amirov
Computer Modelling in Tomography and Ill-Posed Problems
M.M. Lavrent'ev, S.M. Zerkal and O.E. Trofimov
An Introduction to Identification Problems via Functional Analysis
A. Lorenzi
Coefficient Inverse Problems for Parabolic Type Equations and Their Application
P.G. Danilaev
Inverse Problems for Kinetic and Other Evolution Equations
Yu.E. Anikonov
Inverse Problems of Wave Processes
A.S. Blagoveshchenskii
Uniqueness Problems for Degenerating Equations and Nonclassical Problems
S.P. Shishatskii, A. Asanov and E.R. Atamanov
Uniqueness Questions in Reconstruction of Multidimensional Tomography-Type Projection Data
V.P. Golubyatnikov
Monte Carlo Method for Solving Inverse Problems of Radiation Transfer
V.S. Antyufeev
Introduction to the Theory of Inverse Problems
A.L. Bukhgeim
Identification Problems of Wave Phenomena - Theory and Numerics
S.I. Kabanikhin and A. Lorenzi
Inverse Problems of Electromagnetic Geophysical Fields
P.S. Martyshko
Composite Type Equations and Inverse Problems
A.I. Kozhanov
Inverse Problems of Vibrational Spectroscopy
A.G. Yagola, I.V. Kochikov, G.M. Kuramshina and Yu.A. Pentin
Elements of the Theory of Inverse Problems
A.M. Denisov
Volterra Equations and Inverse Problems
A.L. Bughgeim
Small Parameter Method in Multidimensional Inverse Problems
A.S. Barashkov
Regularization, Uniqueness and Existence of Volterra Equations of the First Kind
A. Asanov
Methods for Solution of Nonlinear Operator Equations
V.P. Tanana
Inverse and Ill-Posed Sources Problems
Yu.E. Anikonov, B.A. Bubnov and G.N. Erokhin
Methods for Solving Operator Equations
V.P. Tanana
Nonclassical and Inverse Problems for Pseudoparabolic Equations
A. Asanov and E.R. Atamanov
Formulas in Inverse and Ill-Posed Problems
Yu.E. Anikonov
Inverse Logarithmic Potential Problem
V.G. Cherednichenko
Multidimensional Inverse and Ill-Posed Problems for Differential Equations
Yu.E. Anikonov
Ill-Posed Problems with A Priori Information
V.V. Vasin and A.L. Ageev
Integral Geometry of Tensor Fields
V.A. Sharafutdinov
Inverse Problems for Maxwell's Equations
V.G. Romanov and S.I. Kabanikhin

INVERSE AND ILL-POSED PROBLEMS SERIES

Forward and Inverse Problems for Hyperbolic, Elliptic, and Mixed Type Equations

A.G. Megrabov

///VSP///

UTRECHT • BOSTON

2003

VSP
P.O. Box 346
3700 AH Zeist
The Netherlands

Tel: +31 30 692 5790
Fax: +31 30 693 2081
vsppub@compuserve.com
www.vsppub.com

First published in 2003

ISBN 90-6764-379-3

Printed in The Netherlands by Ridderprint bv, Ridderkerk.

To my dearly loved
and incomparable
sister Emma

Preface

Inverse problems is the important and intensively developing direction in mathematics; namely, in mathematical physics, in differential equations and in various applied technologies: in geophysics, optic, tomography, remote sensing, radar-location etc.

In this monograph the direct and inverse problems for partial differential equations are considered. The type of equations we consider may be hyperbolic, elliptic, and mixed (elliptic-hyperbolic). The direct problems arise as generalization of problems of scattering the plane elastic or acoustic waves from the inhomogeneous layer (or from the half-space). The inverse problems are the problems of determination of the medium parameters by giving the forms of incident and reflected waves or by giving the vibrations of certain points of the medium. The way of determination of the density $\rho(z)$ and the speed of wave propagation $v(z)$ is given. Here $v(z)$ and $\rho(z)$ are the functions of the depth z and we know the family of plane waves reflected from the layer at various angles. This way is realized both analytically and numerically (the numerical algorithm is obtained). Such statements arise, for example, in seismology and in acoustics. The method of research of all the inverse problems is spectral-analytical. It consists in reducing the considered inverse problems to the known inverse problems for the Sturm—Liouville equation or for the string equation.

Besides, in the book, the discrete inverse problems are considered. In the problems we determine an arbitrary set of point sources (emissive sources, oscillators, point masses). The desired values are the form of the impulse (in the nonstationary case), the amplitude, phase, frequency (in the stationary case), the masses (in the static case) of sources. In all the cases we must find the coordinates of the sources and their number, which may be arbitrarily large.

This book may be useful for students and scientists in mathematics, mathematical physics, geophysics and biophysics.

Contents

Introduction

This book is devoted mainly to the inverse problems for differential equations of mathematical physics.

In accordance with the terminology used in the mathematical literature, the inverse problems for linear differential equations are the problems of determining the coefficients of the differential equation of a given class, or the right-hand side of this equation by giving data on the solution of a certain boundary-value (direct) problem for this equation.

In problems studied traditionally in mathematical physics, coefficients of the equation and its right-hand side are assumed to be known. We need to find the solution of the differential equation in the domain under consideration given the initial and boundary conditions. In the inverse problem theory such settings are called direct (or forward) problems.

If a differential equation describes a physical process (physical field), its coefficients describe the characteristics (parameters) of the physical medium in which the process (field) is considered. The right-hand side of the equation describes the sources of the process (field).

Therefore, from the physical point of view, inverse problems are in determining the physical characteristics (parameters) of the medium and (or) the sources of the physical field by using some information on the physical field (solution of the direct problem). Often, in inverse problems, we need to find these characteristics and (or) sources of the field inside a certain domain, and the information is given only at the boundary of this domain. Direct problems are in finding the physical field in the domain under consideration if the characteristics of the medium and the sources are given.

The reader can be acquainted with the theory of inverse problems in monographs of Lavrentiev, M. M. et al. (1967, 1986), Romanov, V. G. (1972, 1973, 1987), Anikonov, Yu. E. (1978, 1995), Kabanikhin, S. I. (1988), Romanov, V. G. and Kabinikhin, S. I. (1991, 1994), Belishev, M. I.

and Blagoveshchenskii, A. S. (1999), Bukhgeim, A. L. (1999, 2000), Kabanikhin, S. I. and Lorenzi, A. (1999).

The inverse problems for linear second-order partial differential equations can be classified by using the following important features:

1. Type of differential equation for which the inverse problem is formulated.

2. Number of independent variables on which the unknown functions depend.

3. Shape of the domain in the space of independent variables where the differential equation is considered.

First, we describe inverse problems using this classification. In accordance with the first feature all our inverse problems are divided into following classes:

1. problems for hyperbolic equations;

2. problems for elliptic equations;

3. problems for mixed-type (elliptic-hyperbolic) equations, in which the equation is hyperbolic in one part of the domain and elliptic in its other part.

In these three classes, all unknown coefficients are functions of one independent variable. The unknown functions may be arbitrary, but must satisfy certain smoothness conditions, i. e., they must belong to a certain functional class.

Both hyperbolic and elliptic equations are considered in two-dimensional domains of two types: in a half-plane and in a strip. The equations of mixed type are considered in a strip.

Various inverse problems are formulated for each class of this classification. They differ by the character of the initial data (the given information about the solution of the direct boundary-value problem).

Unknown characteristics in these problems are, first of all, the coefficients of the differential equation in question; these problems are considered in Chapters 1–7.

In Chapter 8, we consider problems in which an arbitrary set of point sources is unknown. We must find the number of sources, their coordinates, and the physical characteristics of these sources. From the mathematical point of view, in these problems we must find the right-hand side of special form in the wave equation or in the Poisson equation. This right-hand side corresponds to a certain set of point sources of the field.

The main purpose of our book is systematic presentation of the results published by the author (Alekseev, A. S. and Megrabov, A. G., 1972, 1973a,b,c, 1974; Megrabov, A. G. 1972, 1973a,b; 1974, 1975a,b; 1977a,b, 1978a,b).

An important aspect of theoretical investigation of any inverse problem is the proof of the uniqueness theorem. In each inverse problem we formulate and prove this theorem. In each case the method for the proof of the uniqueness theorem is such that it gives the method for its solution. This means that the method is constructive.

The uniqueness theorems obtained and methods proposed to solve the inverse problems form the essence of the monograph.

It should be noted that the boundary-value (direct) problems for equations of all three types are also of interest for specialists in mathematical physics, since these problems are nonclassical. In particular, this remark is relates to the direct problems for equations of elliptic and mixed types. These problems are non-classical first of all owing to one of the boundary conditions. Namely, we mean the values of directional derivative, which we set at one of the boundaries. (In the classical boundary conditions, either the solution of the direct problem, its normal derivative, or a linear combination of these values is given). For each of these direct problems we prove the corresponding uniqueness theorem. Besides, to solve each direct problem, a special integral representation is constructed, which may be of a certain interest.

The first part of the issues considered is related to the inverse problems for the hyperbolic equation of vibrations of an inhomogeneous string, and is considered in Chapters 1 and 4.

Inverse problems for the string refer to the most developed directions in the inverse problem theory. The first fundamental theorems devoted to the inverse problems for the string and the inverse problems for the Sturm—Liouville equation were proved in the classical papers of Ambartsumjan, W. A. (1929), Krein, M. G. (1951a,b, 1952, 1953); Marchenko, M. A. (1950, 1952); Borg, G. (1945); Gel'fand, I. M. and Levitan, B. M. (1951); Gasymov, M. G. and Levitan, B. M. (1964); Levitan, B.,M. (1962); Chudov, L. A. (1949). The important formula which expresses the spectral function of the Sturm—Liouville operator via the solution of a certain integral equation was obtained by Alekseev, A. S. (1962) when solving the inverse problem for the wave equation.

These results are basic for the settings and investigation of the inverse problems considered in Chapters 1–7.

Later, Blagoveshchenskii, A. S. (1971) gave a new proof of the results obtained by Krein, M. G. (1954) without the use of spectral theory. Earlier, a similar approach was developed by Parijskii, B. S. (1969). The problems of uniqueness and stability of the inverse problem for a string with an unknown (under certain conditions) and arbitrary source of vibrations were studied by Gerver, M. L. (1970). Some new interesting theorems for the inverse problems for the string equation were obtained by Blagoveshchenskii, A. S. (1970a,b); Blagoveshchenskii, A. S. and Buzdin, A. A. (1972); Blagoveshchenskii, A. S. and Voevodskii, K. E. (1981); Belishev, M. I. (1975, 1977, 1978); Belishev, M. I. and Blagoveshchenskii, A. S. (1999); Kabanikhin, S. I. (1979, 1984, 1988). For the wave equation and the similar equations, in particular, for the telegraph equation, the inverse problems were considered by Alekseev, A. S. (1962, 1967), Blagoveshchenskii, A. S. (1970a,b); Kabanikhin, S. I. (1988), Belishev, M. I. and Blagoveshchenskii, A. S. (1999); Blagoveshchenskii, A. S. and Voevodskii, K. E. (1981), and by Romanov, V. G. (1972a,b; 1984). The reader can be acquainted with these results in the generalizing monographs of Romanov, V. G. (1984); Lavrentiev, M. M., Romanov, V. G., and Shishatskii, S. P. (1980), Bukhgeim, A. L. (1983, 1987); Kabanikhin, S. I. (1988); Romanov, V. G. and Kabanikhin, S. I. (1991, 1994); Kabanikhin, S. I. and Lorenzi, A. (1999); Belishev, M. I. and Blagoveshchenskii, A. S. (1999), and Anikonov, Yu. E. (1978, 1995).

We shall not discuss these results since they do not relate directly to our book.

All the results obtained by other authors we use in our monograph are formulated in those places of the text where they are applied.

Here only the following important point should be mentioned. All the inverse problems investigated for the string vibration equation (Blagoveshchenskii, A. S., 1966, 1971; Blagoveshchenskii, A. S. and Buzdin, A. A., 1972; Gerver, M. L., 1970; Krein, M. G., 1951a,b; 1952; 1953; 1954; Parijskii, B. S., 1969; Romanov, V. G., 1973; 1987; Kabanikhin, S. I., 1979; 1984; 1988; Belishev, M. I., 1975; 1978; Belishev, M. I. and Blagoveshchenskii, A. S., 1999) are related to the case when boundary conditions of the form

$$\left(\beta\,\frac{\partial u}{\partial z}+\alpha u\right)\Big|_{z=\mathrm{const}}=f(t)\qquad(-\infty<\alpha,\beta<\infty)\tag{0.0.1}$$

are given at the ends of the string.

From the physical point of view, these inverse problems are in finding mechanical properties of the string by the motion of its certain points.

These problems are connected directly with the inverse Sturm—Liouville problem (see, Gasymov, M. G. and Levitan, B. M., 1964; Gel'fand, I. M. and Levitan, B. M., 1951; Krein, M. G., 1951a,b; 1952; 1953; 1954; Levitan, B. M., 1962; Levitan, B. M. and Sargsyan, I. S., 1970; Marchenko, V. A., 1950; 1952; Chudov, L. A., 1949; Ambartsumjan, W. A., 1929; Borg, G., 1945), since a self-adjoint one-dimensional differential operator with real constant coefficients in the boundary conditions corresponds to these boundary conditions. All important notions such as the spectral function (Gel'fand, I. M. and Levitan, B. M., 1951; Levitan, B. M. and Sargsyan, I. S., 1970), the main transition function (Krein, M. G., 1953; 1954), and main theorems from the theory of inverse Sturm—Liouville problems refer to this operator.

In our monograph, the inverse problems of determining the variable coefficients in the hyperbolic equation of the string vibrations

$$\frac{\partial}{\partial z}\Big\{\mu(z)\,\frac{\partial u}{\partial z}\Big\}=\eta(z)\,\frac{\partial^2 u}{\partial t^2}\tag{0.0.2}$$

are considered. The string is considered both semi-bounded and finite, and at one end of the string the following boundary condition of the directional derivative

$$\left(\frac{\partial u}{\partial z}-\varkappa\,\frac{\partial u}{\partial t}\right)\Big|_{z=\mathrm{const}}=f(t)\qquad(\varkappa\geq 0),\tag{0.0.3}$$

is given. This condition means that, in addition to constraining force, a force proportional to the speed of motion acts on this end. Such inverse problems have not been investigated before. However, some seismology and hydroacoustic problems lead to such problems; therefore, they are of interest for our investigation.

An essential difference of the problem with conditions of type (0.0.3) from the problem with conditions of type (0.0.1) is that the corresponding one-dimensional operator is (for $\varkappa\neq 0$) not self-adjoint, with a complex variable coefficient in the boundary condition. So, to obtain the theorems presented in Chapters 1 and 4, we need to use the results of Naymark, M. A.

(1954, 1969) in the theory of not self-adjoint differential operator of the second order. We have reduced all inverse theorems with condition (0.0.3) to the known inverse Sturm—Liouville problem, which is formulated for the self-adjoint operator. Namely, the inverse problems for the semi-bounded string (half-plane) were reduced to the problem of determining the Sturm—Liouville operator in the half-axis by using the spectral function or main transition function. All problems for a finite string were reduced to the problem of determining the Sturm—Liouville regular operator given two spectra (see Gasymov, M. G. and Levitan, B. M., 1964; Krein, M. G., 1951; Levitan, B. M., 1962; Marchenko, V. A., 1952). This reduction is achieved by using the following method. In the latter case, it is as follows. First, we consider, parallel with our not self-adjoint differential system, two self-adjoint regular Sturm—Liouville problems. Then, using the explicit representation of the solution of the direct problem, we find the spectra of these two problems.

This approach allows us to prove the uniqueness theorems and suggest a way of solving the inverse problems for the string with the condition of the directional derivative at one end. This is considered in Chapters 1 and 4.

The second part of the results refers to the theory of inverse problems, where the coefficients in elliptic equations are to be found. These problems are considered in Chapters 2 and 5.

This area has been poorly studied[i)]. The well-known result obtained in the monograph of Lavrentiev, M. M. et al. (1967) is formulated as follows. Denote by $G(\lambda, P, Q)$ the Green function of the first (second) kind of the equation

$$\Delta v = (a^2 + \lambda b(P))v, \quad (a^2 + \lambda b(P)) \geq 0, \quad P = (z, \xi) \tag{0.0.4}$$

considered in the half-plane $z > 0$, where $b(P) \equiv 0$ for $z < z_1 = \text{const} > 0$.

Here a^2 is a constant, λ is a parameter, and b is a bounded continuous function. The inverse problem: find the coefficient $b(z, \xi)$ if the function

$$f(\xi, \xi_0) = \frac{\partial}{\partial \lambda} G(\lambda, P, Q)\Big|_{P=(z_1,\xi),\, Q=(z_1,\xi_0)}^{\lambda=0} \tag{0.0.4$'$}$$

is given for $-\infty < \xi, \xi_0 < \infty$ — may have not more than one solution in the class of absolutely integrable functions.

[i)]As was noted by Lavrentiev, M. M. et al. (1967), inverse problems of potential theory, in which the right-hand side of the elliptic equation must be determined, have been investigated more good. As for the determination of the coefficients of elliptic equations, the theory of these inverse problems has not been completely developed yet.

Thus, the data of the inverse problem are multidimensional: the derivative of the function $C(\lambda, P, Q)$ with respect to the parameter λ is given, as a function of the parameter ξ_0 and the point ξ of the straight line $z = z_1$. However, the inverse problem is also multidimensional: $b = b(z, \xi)$.

Other inverse problems formulated for elliptic equations and developed later are considered by Romanov, V. G. (1984) (here see Chapter 6 and Bibliography). The inverse problems for quasilinear equations of elliptic type were investigated by Iskenderov, A. D. (1972) and, for the Helmholtz equation, by Zapreev, A. S. and Tsetsokho, V. A. (1976). An interesting approach was suggested by Sylvester, J. and Uhlmann, G. (1987). It is based on the Dirichlet—Neumann map. In this case, the inverse problem is formulated in a highly overdetermined statement, when an infinite number of functions of a certain type is given in the boundary condition.

Nevertheless, it seems that prior to the author's publications (see Megrabov, A. G., 1972; 1973b,c; 1974, 1975a), there were no results on the inverse problems (even one-dimensional) of determination of the coefficients in the elliptic equations, in which, as the initial data, the solution of the boundary value problem was given as a function of the boundary points only. In other words, no inverse problems for elliptic equations were considered similar to those considered for the hyperbolic equation of string vibrations (Blagoveshchenskii, A. S., 1971; Krein, M. G., 1954; Romanov, V. G., 1973). In our monograph we consider the results in this direction.

The necessity of consideration of such problems has two reasons.

First, as we show below, there exist some important applications that lead to the inverse problems for elliptic equations on determining the coefficients when the initial data is given as the solution of the boundary-value problem at the boundary of the domain. This solution is given on the same way as the vibration of one string end (the main transition function) for the determination of string density (the coefficient in the hyperbolic equation).

Second, the elliptic equation, in this case, has a form different from (0.0.4):

$$\frac{\partial^2 u}{\partial z^2} + \eta(z)\,\frac{\partial^2 u}{\partial \xi^2} + \tau(z)\,\frac{\partial u}{\partial z} = 0 \qquad (0 < m < \eta(z) < M < \infty). \qquad (0.0.5)$$

It does not contain the parameter λ; therefore, the inverse problem with the data of type (0.0.4′) cannot be formulated.

In our monograph (Chapters 2 and 5), inverse problems on the determination of the coefficients in elliptical equation (0.0.5) are considered in a half-plane or a strip. The data is the solution of the boundary-value (direct) problem for this equation given for the boundary points only.

Thus, inverse problems for elliptic equations considered in our book differ from the problem formulated by Lavrentiev, M. M. et al. (1969) in the form of equation and the inverse problem data. Besides, in addition to the inverse problems considered in the half-plane, we have considered the case when the elliptic equation is given in the strip.

The basic results of our book devoted to the inverse problems for elliptic equations are the uniqueness theorems formulated in Chapters 2 and 5.

If one compares these theorems with theorems of Chapters 1 and 4, he will see the analogy between the inverse problems for elliptic equation (0.0.5) and the problems formulated for the hyperbolic equation of string vibration (0.0.2) from Chapters 1 and 4. This is so because the method for solving both the inverse problems for elliptic equations and hyperbolic equations is in their reduction to known inverse problems for the Sturm—Liouville equation, in spite of the fact that the correspondent one-dimensional operator is not self-adjoint and not a Sturm—Liouville operator. Namely, all the inverse problems considered in the half-plane are reduced to the determination of the Sturm—Liouville operator in the semiaxis by using the main transition function (Krein, M. G., 1953; 1954) or the spectral function (Gel'fand, I. M. and Levitan, B. M., 1951; Marchenko, V. A., 1952; Gasymov, M. G. and Levitan, B. M., 1964). All inverse problems considered in the strip are reduced to the problem of determining the regular Sturm—Liouville operator by using two spectra (Gel'fand, I. M. and Levitan, B. M., 1951; Marchenko, V. A., 1952; Levitan, B. M., 1962; Gasymov, M. G. and Levitan, B. M., 1964).

Thus, in spite of essential differences in the inverse problems (equation type, the domain shape, initial data form), the method of solution and the proof of uniqueness theorems are similar: the inverse problem is reduced to the inverse problem for the Sturm—Liouville equation. These inverse problems for the Sturm—Liouville equation can be divided into two classes.

First, this is the well-known problem of determining the real function $q(x)$ and the real number α in the differential system:

$$\begin{cases} -y'' + q(x)y = \lambda y, & x > 0, \\ y'(0) - \alpha y(0) = 0. \end{cases} \tag{0.0.6}$$

The differentiation with respect to x is denoted by a prime. The spectral function of this system or its main transition function are given. All inverse problems considered for hyperbolic and elliptic equations in the half-plane are reduced to this problem.

Second, this is a problem of determining the regular Sturm—Liouville operator by two spectra (Gasymov, M. G. and Levitan, B. M., 1964; Krein, M. G., 1951; Levitan, B. M., 1962; Marchenko, V. A., 1952). It is formulated as follows: given the spectra of two boundary-value problems:

$$\begin{gathered} -y'' + q(x)y = \lambda y, \qquad 0 \le z \le l < \infty, \\ \cos\varphi_i y(0) + \sin\varphi_i y'(0) = 0, \qquad i = 1, 2, \quad \varphi_1 \ne \varphi_2 \\ \cos\theta y(l) + \sin\theta y'(l) = 0 \end{gathered} \tag{0.0.7}$$

with the same boundary condition at one end of the interval and different boundary conditions at the other end. We need to find the real function $q(x)$ and the real numbers l, $\operatorname{ctg}\varphi_1$, $\operatorname{ctg}\varphi_2$, and $\operatorname{ctg}\theta$.

We reduce all inverse problems for elliptic and hyperbolic equations in the strip to this problem.

This approach to each inverse problem under consideration is based on a special representation for the solution of the correspondent direct boundary-value problem. A priori we do not know the form of this representation for each concrete inverse problem. However, the search of a representation convenient for the solution of the inverse problem forms essential part of our work.

The similar approach to the solution of the inverse problems for partial differential equations was applied, for example, by Krein, M. G. (1953, 1954) and Gerver, M. L. (1970) to inverse problems for the string equation and by Alekseev, A. S. (1962, 1967) to inverse problems for the wave equation. In accordance with the terminology of Alekseev, A. S. (1967) and Blagoveshchenskii, A. S. (1971) we call this approach spectral.

In the works mentioned above, inverse problems are formulated for hyperbolic equations with the boundary conditions of type (0.0.1). Therefore, in corresponding one-dimensional problems (obtained from the initial problems after the use of the method of separation of variables), boundary conditions have the form (0.0.6) or (0.0.7) and are self-adjoint. Differential systems have the form (0.0.6) or (0.0.7), that is, they are systems for which the known inverse problems for the Sturm—Liouville equation are formulated.

In our monograph, we extend the area of application of the spectral approach. This extension is done in two directions.

First, in our inverse problems the corresponding one-dimensional operator is not self-adjoint (in contrast to the known works by Alekseev, 1962, 1967, etc). Nevertheless, these problems are reduced to the known inverse problems for self-adjoint differential Sturm—Liouville system of the form (0.0.6) or (0.0.7). Thus, the one-dimensional operator we consider is of a more general form. Note that in all problems considered in our monograph the one-dimensional operator is not self-adjoint because it does not contain a boundary condition of the form (0.0.6) or (0.0.7). It contains a more general boundary condition with a complex variable (depending on the spectral parameter k, $k^2 = \lambda$) coefficient $(\alpha + w(k)\beta)$. Here α and β are numbers, and $w(k)$ is a complex function.

Second, the inverse problems for *elliptic* equations and for *mixed* type equations (Chapter 7) are reduced to the inverse problems for the Sturm—Liouville equation.

The settings of inverse problems for the string and elliptic equations considered in the monograph are not accidental. They had arised as generalization of some inverse dynamic seismic problems (some inverse problems of wave propagation theory).

Inverse dynamic seismic problems is a new direction in geophysics. Its general ideology is formulated by Alekseev, A. S. (1962, 1967). This direction originated because of a tendency to consider a more general structure of the physical media. It is known that the solution of the inverse kinematic problems of seismics is not unique in the case of non-monotone change in the wave propagation speed in space (see Alekseev, A. S., 1967; Romanov, V. G., 1972). The dynamic approach allows us to consider more general properties of the medium. On the other hand, the dynamic approach allows us to give a more complete description of the properties of the medium. Alekseev, A. S. (1962) showed that, on the basis of dynamic seismic problems, we can determine not only the speed of wave propagation, but also the density. We cannot find this characteristic (density) using only kinematic methods, since kinematic methods use only a part of useful information on vibrations of medium's points.

In other words, all considered inverse problems for hyperbolic and elliptic equations are combined both by the investigation method and by the generality of the physical problems that lead to them. All these problems can be called inverse problems of scattering of the plane waves from inhomogeneous layers, since they originated in the following situation.

Suppose that we have an inhomogeneous elastic layer with the following properties: the density ρ and the speed v and these functions dependent on the normal coordinate z (the depth). The two types of media will be considered. In the first type, the inhomogeneous layer is a half-space or, in the particular case, it is a transition layer. The model is considered in

Chapter 3. In the second model, considered in Chapter 6, the inhomogeneous layer has a finite depth H and has a free (or fixed) boundary $z = H$. In both cases, the plane $z = 0$ is the boundary or rigid contact of the inhomogeneous layer with the homogeneous half-space with the parameters ρ_0 and v_0.

In each model the following question arises: can we find the unknown properties of the inhomogeneous layer if we excite the inhomogeneous layer by a plane wave falling obliquely and measure in a certain point the wave reflected from the layer?

Besides, in the case of the layer with free (fixed) boundary the problem arises: to find the characteristics of the inhomogeneous layer by the record of vibrations of a fixed point from the free boundary.

More explicitly, we must show which data are sufficient for unique determination of the properties of the inhomogeneous layer, i. e., we must to formulate the uniqueness theorem.

The third group of the inverse problems is devoted to these problems (Chapters 3 and 6).

We consider the case when the incident plane wave is an elastic transverse wave of the SH-type, or an acoustic wave. The problem with an longitudinal elastic wave obliquely is a more complicated problem. However, in the case of a cross-wave, there arise various inverse problems for differential equations.

Formulate now the inverse problems not investigated before which arise in the first model of the medium (Chapter 3).

One of the inverse problems arised in the theory of wave propagation is the problem of determination of the properties of inhomogeneous elastic layer given the form of incident and reflected plane waves in the case of normal incidence. Various methods and algorithms are suggested by Nikol'skii, E. V. (1964); Borodaeva, N. M. (1969); Mikhailov, N. G. and Parijskii, B. S. (1969). This problem is of interest in connection with the important problem of investigation of marine sediments. The first attempts to apply the solution of this problem to investigate marine sediments are made by Borodaeva, N. M. (1969).

Uniqueness of the solution to this problem was expected; however, a formulation of the uniqueness theorem was absent. In our monograph this theorem is obtained as a particular case of a more general statement.

It is essential that only the acoustic shiftness $\sigma_0(x) = \rho(x)v(x)$ can be obtained (as a function of the wave travel time x to the depth z) given the information that is usually given in the problem of normal incidence of a plane wave. All the papers mentioned above are devoted to the determination of the function $\sigma_0(x)$. The density ρ and the wave propagation speed v in the inhomogeneous layer as functions of the depth z remain unknown.

In this connection the following problems are of interest.

First, more complete determination of the elastic properties of the inhomogeneous layer, for example, the functions $v(z)$ and $\rho(z)$.

Second, the study of inverse problems for the general case of arbitrary oblique incidence of the plane wave.

The general theorems obtained in Chapters 1 and 2 of our monograph allow us to solve the both questions important for the theoretical seismology in the case of incident transverse SH-waves. These problems are connected with each other, since the functions $v(z)$ and $\rho(z)$ are recovered uniquely only when we pass from the particular case of normal incidence to the general case of oblique incidence of a plane wave.

When we pass from normal incidence to a oblique incidence it occurs that the differential equation that contains the unknown coefficients may have hyperbolic, elliptic, or mixed types. The hyperbolic case is defined by the condition $v(z) < v_0/\sin\theta_0$ (the apparent speed is greater than the speed in the layer) and corresponds to subcritical incidence of the plane wave. In particular, the normal incidence refers to this case.

A quantatively new problem connected with another type of differential equation in comparison with the normal incidence arises in the elliptic case: $v(z) > v_0/\sin\theta_0$. In physics, it is called the case of complete internal reflection (Brekhovskih, L. M., 1973), or supercritical incidence. Inverse problems for this case were not considered before the works of Megrabov, A. G. (1972, 1973a,b,c, 1974, 1975).

In both cases, first of all, we formulate the problem of determining the properties of the inhomogeneous layer by the forms of incident and reflected waves for one value θ_0 of the angle of incidence of the plane wave. The uniqueness theorem and the way of its solution are obtained by a simple transfer of the results of Chapters 1 and 2. It occurs that in the general case we can recover uniquely a certain intermediate characteristic of the layer, the function $\sigma(x,\theta_0)$ (which coincides, at $\theta_0 = 0$, with the function $\sigma_0(x)$ mentioned above). The functions $v(z)$ and $\rho(z)$ are not determined.

In this connection, in Chapter 3 the method of recovery of the density and speed of wave propagation in the inhomogeneous layer as functions of the depth is suggested. A set of plane waves reflected from the layer at various angles is given. The functions $v(z)$ and $\rho(z)$ are recovered uniquely both in the hyperbolic and elliptic cases by specifying the forms $\{\varphi_0(\xi,\theta_0);\varphi_1(\xi,\theta_0)\}$ of incident and reflected waves for a certain set of values θ_0. The uniqueness theorem for this inverse problem is proved in § 3.3. The first inverse problem with the data given for one value θ_0 can be considered as an independent problem which generalizes, in the hyperbolic case, the problem of normal incidence. However, with respect to the problem of determining the functions $v(z)$ and $\rho(z)$, it is auxiliary.

In the monograph (Chapter 3), a numerical algorithm for the determination of the density and speed of wave propagation as functions of the depth is suggested. The functions $v(z)$ and $\rho(z)$ are derived numerically, and the results for the hyperbolic case are given. As was noted, the results obtained before (Nikol'skii, E. V., 1964, etc) are for the intermediate characteristic $\sigma_0(x)$. In our case, it suffices to specify the forms of reflected and incident waves for two or three values of θ_0 in order to recover the functions $v(z)$ and $\rho(z)$ numerically.

Numerical experiments are made to determine the efficiency of the algorithm, i. e., the accuracy and depth of determining $v(z)$, $\rho(z)$, and the stability of the solution of the inverse problem with respect to noise in the inverse problem initial data. In the control examples, by three calculated and set theoretical seismograms $\varphi_1(\xi, \theta_0)$, the functions $v(z)$ and $\rho(z)$ were recovered to a depth of 10 km at 1000 points with a spacing of 10 m. The maximal relative errors in determination of $\rho(z)$ and $v(z)$ were not greater than 3–5% at the absence of noise in the theoretical seismograms.

The results of Chapter 3 were considered briefly by Alekseev, A. S. and Megrabov, A. G. (1972, 1973, 1974), and Megrabov, A. G. (1972, 1974, 1975). Later, Blagoveshchenskii, A. S. and Voevodskii, K. E. (1981) investigated the inverse problems of scattering of plane waves from an inhomogeneous half-space in the hyperbolic case for the differential equation of a more general form.

All this was related to the inverse problems formulated for the first model of the medium. Now, we consider the inverse problems of scattering a plane waves from inhomogeneous layers with a free or fixed boundary (Chapter 6).

Prior to the papers of Alekseev, A. S. and Megrabov, A. G. (1974) and Megrabov, A. G. (1973b,c; 1974, 1975) there were no results on this problem. Nevertheless, these problems are of interest in connection with the following seismology problem formulated by Alekseev, A. S. (see Lavrentiev, M. M., 1972).

Let us have a plane elastic layer bounded from below by a homogeneous elastic medium and from above by a free boundary. The characteristics of the layer v_p, v_s, and ρ depend on the vertical coordinate z only. At the free surface, the result of interaction of the layer with a set of oblique plane waves incoming from infinity is recorded. The functions v_p, v_s, and ρ are to be found. The solution of this problem can be used for determining the structure of the Earth's crust by the recording of distant earthquakes at seismic stations. Thus, the inhomogeneous layer is a model of the Earth's crust, and the free boundary corresponds to the day surface.

The uniqueness theorem for this problem is proved in Chapter 6. A method for its solution is suggested in the case of scattering of SH-waves. Here, the three cases (hyperbolic, elliptic and mixed) are distinguished as well.

Since no results were obtained for the inverse problems of scattering of plane waves in the inhomogeneous layer with a free boundary, we should know what properties of the layer can be determined if the vibration of a certain point from the free boundary and the form of the incident wave for an arbitrary value of θ_0 are specified. From the general theorems of Chapters 4 and 5 it follows that, in the general case, only a certain function $\sigma(x, \theta_0)$ (coinciding with the function $\sigma_0(x) = \rho(x)v(x)$ for $\theta_0 = 0$) can be determined. The functions $v(z)$ and $\rho(z)$ remain unknown. Therefore, we need to increase the information specified (see Lavrentiev, M. M., 1972).

In the monograph, it is shown that it suffices to take a sequence of values θ_0 that converges to a limiting point, and the recording of vibrations of a fixed point of the free boundary is given for all such θ_0. In this case, we define the functions $\rho(z)$, $v(z)$, and the form $\varphi_0(\xi, \theta_0)$ of the incident wave uniquely in the hyperbolic and elliptic cases. Besides, if the form of incident wave is known, there exists the way of recovery of $v(z)$, $\rho(z)$ and H by the vibrations of a point of the free boundary given for a certain set of angles θ_0. In the numerical variant, it suffices to have two or three values of θ_0. This method is similar to that suggested in Chapter 3. In Chapter 6, the characteristics $\sigma(x, \theta_0)$, $v(z)$ and $\rho(z)$ in the inhomogeneous layer with a free boundary are determined also by the forms of incident and reflected waves.

The fourth part of the problems in question is related to the direct and inverse problems for equations of mixed (elliptico-hyperbolic) type and is considered in Chapter 7. These problems are published in brief form by the author (Megrabov, A. G., 1975, 1977a, 1977b). It should be noted that prior to these investigations, apparently, there were no studies devoted to the inverse problems for equations of mixed type. Therefore, these inverse problems are the first results obtained for differential equations of mixed type.

We consider the equation of mixed type

$$K(z)\,\frac{\partial^2 u}{\partial \xi^2} + \frac{\partial^2 u}{\partial z^2} = 0$$

in the strip $0 < z < H$, $-\infty < \xi < \infty$, where

$$K(z) < 0 \quad \text{for} \quad z \in [0, h)$$

and

$$K(z) > 0 \quad \text{for} \quad z \in (h, H],$$

$0 < h < H < \infty$. In the particular case, when

$$K(z) = \operatorname{sign}(z - h) = \pm 1,$$

we obtain the well-known Lavrentiev–Bitsadze equation.

We admit any combination of the conditions

$$K(h + 0) = 0, \quad K(h + 0) \neq 0, \quad K(h - 0) = 0, \quad K(h - 0) \neq 0.$$

This means that the coefficient $K(z)$ in the equation when passing from the hyperbolicity domain to ellipticity domain may change both continuously (vanish in the transition line $z = h$) and discontinuously (have a jump). In the continuous case, the power of vanishing at the point $z = h$ may be arbitrary and different from the both sides. In the case of the jump, $K(z)$ may vanish only from one side of the point $z = h$, or not vanish at all.

Similar to the problems considered in Chapters 1–6 for the equations of hyperbolic and elliptic types we use the spectral-analytic method; and the inverse problems of determining the coefficient $K(z)$ for our equation of mixed type are reduced to the well-known inverse problems for the Sturm—Liouville equation or to the string equation. For this purpose, the integral representation obtained for the solution of the direct problem is used. At first, the direct problem is formulated correctly. In § 1, the uniqueness of its solution is proved, since the formulation of the direct problem that we consider is absent in the available literature for equations of mixed type (see, for example, Smirnov, M. M., 1970). (Our statement of the direct problem has a condition of the directional derivative in one of the boundary components of the strip).

Note that the proof complexity of the solution representation for the direct problem depends on the fact whether the coefficient $K(z)$ vanishes at least on one side of the point $z = h$ or is different from zero from each of the sides of this point. The last particular case when $K(h + 0) \neq 0$, $K(h - 0) \neq 0$ published in the paper by Megrabov, A. G. (1975, 1977b) is considered separately in § 2, 3, as a relatively simple one. The general case of behaviour of the coefficient $K(z)$ in the transition line, which includes all particular cases, uses asymptotic formulas of Langer, R. E. (1931). This case is stated in § 5 and is published briefly in Megrabov, A. G. (1977a).

It seems that not only inverse problems but the direct problem formulated in § 1 and the representation obtained for the solution of the direct problem (§ 2–5) are new. These results are of interest for the theory of boundary-value (direct) problems for equations of mixed type, since these settings are nonclassical.

Note that the known inverse problems of determination of the coefficient in the equation of string vibrations are interpreted as problems of determination of its physical characteristics: variable nonnegative density and the length of the string, by given vibrations of its ends. The inverse problems for equations of mixed type can be interpreted in a similar way. These problems can be interpreted as problems of determination of the parameters of a string whose density may have both positive and negative values in various segments of the string.

The origination of these inverse and direct problems for equations of mixed type and for the hyperbolic and elliptic equations in Chapters 1–6 is also connected with the problem of scattering the plane waves from the inhomogeneous layers. The case of the equation of mixed type arises when in one part of the layer $v(z) < v_0/\sin\theta_0$ (i. e. the apparent speed is greater than the speed of wave propagation in the layer; this is the subcritical incidence of the plane wave), and in the other part of the layer $v(z) > v_0/\sin\theta_0$ (i. e. the apparent speed is less than the speed of wave propagation in the layer; this is the so-called supercritical incidence of the plane wave; this case corresponds to the case of complete internal reflection).

Later, Belishev, M. I. (1977, 1978) considered a more general case in the direct and inverse problems of scattering of plane waves from an inhomogeneous half-space and in the corresponding problem for the equation of mixed type. This was the case when the coefficient $K(z)$ may change its sign not once, but without vanishing at the points where the sign changes. So $K(z)$ has jumps at these points. Note also that in these problems of scattering the plane waves considered in Chapter 7, the scattering medium is not an inhomogeneous half-space, but an inhomogeneous layer with a free or fixed boundary.

The fifth part of the book represents inverse problems connected with the determination of an arbitrary set of point sources (Chapter 8). These results are published in brief form in Megrabov, A. G. (1977b; 1978b).

These inverse problems arise in connection with the following problem (1977b). Suppose that in a half-space, or in the whole space, N point sources are located. These sources generate a summary field u in the medium, which satisfies the wave equation. The question which arises is whether there exists a set E of points at the boundary of the half-space (in the case of the space — on an arbitrary plane such that all the sources are situated from one side of this plane), such that the specification of the field u at the points of E allows us to determine uniquely the unknown quantities. In the nonstationary case, it is the number N, and the coordinates and the form of the impulse of each source; in the stationary case, it is the number N, the coordinates, amplitude, phase, and the frequency of each source. The number N in both cases may be arbitrarily large.

In Chapter 8, it is shown that this set E exists. In the case when N is unknown, E may be, in particular, an arbitrary bounded domain or a certain countable set of points. In the case when N is also unknown but it is known that $N \leq n$, where n is known, we show that the set E consists of a finite number of points depending on n. In particular, the set E may be taken as the following set consisting of $\{(C_{2n}^2 + 1)f(2n)\}$ points: in the plane, we take $(C_{2n}^2 + 1)$ arbitrary straight lines, and in each line we take $f(2n)$ arbitrary points. Here C_n^m is a binomial coefficient, and $f(n) = n\{n(n^2 - 1)/2 + 1\}$. Similar results are obtained for the static problem. In this problem, we should find the number N, the coordinates and the masses (generally speaking, the masses may have the both signs) of an arbitrary set of point sources of the Newton potential.

The problems we consider in this chapter are problems of determining the right-hand side of a special form in the wave equation or in the Poisson equation. Their setting is close to the inverse potential problems (Prilepko, A. I., 1973) and to those considered in Zapreev, A. S. (1976), Kardakov, V. B. (1976), and Bleistein, N. and Cohen, J. K. (1977). Later, the inverse problems consisting in the determination of point sources in asymptotic setting within the framework of the antenna theory were considered by Bukhgeim, A. L. et al. (1985, 1995).

The problems we consider in this chapter are of interest in situations when we are dealing with a set of point sources of the wave field of vibrations (mechanical, acoustic or electromagnetic) or a static field (for example, the gravity field). We do not know the field of each source and we cannot measure the coordinates of the sources. We can measure only the summary field, and it is important to find the parameters of each point source.

This situation arises in some geophysical problems, in acoustics, and in bioacoustics. In particular, such problems arise in the problem of sound language of living organisms (for example, of bees, Es'kov, E. K., 1977).

A similar situation arises in biophysics in the problem of information value of some types of biological radiation (see, Prosser, L. and Brown, F., 1967; Tarusov, B. N. et al., 1967; Gurvich, A. A., 1968; Kaznacheev, V. P. et al., 1974; Chumakova, R. I. and Gitel'zon, I. I., 1975). In this case, the problems considered in Chapter 8 may be considered as the simplest mathematical model of "inverse problems of bioluminescence" (in a wide sense). These problems are considered in detail in § 8.12.

This work was performed at the Laboratory of Mathematical Problems of Seismology of the Institute of Computational Mathematics and Mathematical Geophysics, Siberian Branch of Russian Academy of Sciences (formerly Computing Center of Siberian Branch of USSR Academy of Sciences).

The results of Chapter 1 and Chapter 3, § 1, 2 were obtained together with academician Alekseev, A. S. (see, Alekseev, A. S. and Megrabov, A. G., 1972, 1973, 1974). These results are included in order to show, in a more complete form how similar methods can be used to investigate the inverse problems for differential equations of various types (and various inverse problems for the same equation).

I wish to thank my teacher academician Anatolii Semenovich Alekseev for fruitful discussions, remarks, and valuable advises, and continuous support my work.

It should also like to thank to Professor Kabanikhin, S. I. for his interest in my work and support and to Professor Bukhgeim, A. L. for fruitful discussions.

I thank also Professor Belinskii, S. P. and Judina, O. A. for their helpful cooperation in translating the book.

Chapter 1.

Inverse problems for semibounded string with the directional derivative condition given in the end

In this chapter we shall prove the uniqueness theorems for two coefficient inverse problems considered for the hyperbolic equation

$$\frac{\partial}{\partial z}\left\{\mu(z)\,\frac{\partial u}{\partial z}\right\} = \eta(z)\,\frac{\partial^2 u}{\partial \xi^2}, \qquad 0 < z < \infty, \quad -\infty < \xi < \infty, \tag{1.0.1}$$

which describes the semibounded string vibration. In the end of the string the directional derivative condition is given:

$$\left\{\frac{\partial u}{\partial z} - \varkappa\,\frac{\partial u}{\partial \xi}\right\}\Big|_{z=0} = f(\xi). \tag{1.0.2}$$

The method of the proof is constructive. The first inverse problem has the meaning both for $\varkappa \neq 0$ and for the case $\varkappa = 0$ researched before. The second inverse problem has no analog in the case $\varkappa = 0$.

As it was mentioned in the Introduction, the important case $\varkappa = 0$ was researched in the classical works of Krein (1951, 1954), Blagoveshchenskii (1966, 1971), Gerver (1970) and is corresponded to the self-adjoint one-dimensional Sturm—Liouville system. When $\varkappa \neq 0$, the one-dimensional operator is not self-adjoint and Sturm—Liouville. More precisely, we obtain the differential system with the complex (even variable) coefficient in the boundary condition. Nevertheless, we succeed to reduce these two problems to the known inverse Sturm—Liouville problem related to the self-adjoint operator. In order to do this we consider a certain self-adjoint Sturm—Liouville system. Then we determine spectral or main transition function of this system. (In the second case, thus, the inverse problems for the string with condition (1.0.2) for $\varkappa \neq 0$ is reduced to the known inverse problem for the string with condition (1.0.2) for $\varkappa = 0$). Such method is based on using the special representation found for solution of boundary (direct) problem and the formula of A. S. Alekseev for spectral function. We establish this representation using the properties of not self-adjoint operator of the second order in the semiaxis (these properties are established by Naymark (1954)). We consider the sets of such operators and apply the results of Alekseev (1962).

1.1. FORMULATION OF THE DIRECT PROBLEM

Boundary-value (direct) problem 1.1. We consider in the domain $z > 0$, $-\infty < \xi < \infty$ the equation

$$\frac{\partial}{\partial z}\left\{\mu(z)\,\frac{\partial u}{\partial z}\right\} = \eta(z)\,\frac{\partial^2 u}{\partial \xi^2} \qquad (\eta(z) > \eta_0 > 0,\ \ \mu(z) > 0); \tag{1.1.1}$$

the boundary condition with the directional derivative

$$\left(\frac{\partial u}{\partial z} - \varkappa\,\frac{\partial u}{\partial \xi}\right)\Big|_{z=0} = f(\xi) \qquad \left(f(\xi)\big|_{\xi<0} = 0\right); \tag{1.1.2}$$

and the initial condition

$$u(z,\xi)\big|_{\xi<0} = 0, \qquad z \geq 0. \tag{1.1.3}$$

We suppose also that the following conditions hold.

The function $u(z,\xi)$ and the derivative u_ξ are continuous in the half-plane $z \geq 0$, $-\infty < \xi < \infty$ and the derivatives u_{zz}, $u_{\xi\xi}$, $u_{z\xi}$ are continuous in the quadrant $z \geq 0$, $\xi \geq 0$.

One of the functions u_z and u_ξ (any of them) tends to zero uniformly relatively ξ in each finite interval for $z \to \infty$, the other is bounded for $z \geq 0$, $\xi \geq 0$.

The functions $\eta(z)$, $\mu(z)$, and $\partial\mu/\partial z$ are continuous in the interval $0 \leq z < \infty$; the number $\varkappa$ is nonnegative: $\varkappa \geq 0$.

We need to find the function $u(z,\xi)$ defined in the half-plane $z \geq 0$, $-\infty < \xi < \infty$ satisfying all the above conditions.

This problem we shall call the boundary-value direct problem 1.1. Unlike the inverse problems (§ 1.3, 1.4) the number $\varkappa \geq 0$ and the functions $\eta(z)$, $\mu(z)$ $(0 \leq z < \infty)$, $f(\xi)$ $(-\infty < \xi < \infty)$ in direct problem 1.1 are given.

Statement. *The solution of boundary-value problem 1.1 is unique.*

Proof. We suppose that there exist two solutions $u_1(z,\xi)$ and $u_2(z,\xi)$ of the problem and denote their difference by $u(z,\xi) = u_1(z,\xi) - u_2(z,\xi)$. The function $u(z,\xi)$, evidently, satisfies equation (1.1.1), homogeneous condition (1.1.2) (with zero in the right-hand side) and all remained conditions of the boundary-value problem 1.1.

We consider the expression

$$E(\xi) = \frac{1}{2}\int_0^l \{\mu(u_z)^2 + \eta(u_\xi)^2\}\,\mathrm{d}z.$$

The physical sense of the function $E(\xi)$ is that it is the complete energy of the string interval $0 \leq z \leq l$ in the time moment ξ. Differentiating $E(\xi)$ with respect to ξ we differentiate under the integral sign (for differentiating under the integral sign it suffices that the obtained integrand expression be continuous in the interval $0 \leq z \leq l$ for $\xi \geq 0$. In our case this holds true since the functions $u(z,\xi)$, $\eta(z)$, $\eta(\xi)$ are *a priori* sufficiently smooth) we obtain

$$\frac{\mathrm{d}E(\xi)}{\mathrm{d}\xi} = \int_0^l \{\mu u_z u_{z\xi} + \eta u_\xi u_{\xi\xi}\}\,\mathrm{d}z.$$

Integrating by parts the first term of right-hand side we obtain

$$\frac{\mathrm{d}E(\xi)}{\mathrm{d}\xi} = \int_0^l u_\xi\{\eta u_{\xi\xi} - (\mu u_z)_z\}\,\mathrm{d}z + \mu u_z u_\xi\Big|_{z=0}^{z=l}.$$

Integrating with respect to ξ from 0 up to ξ, taking into account equation (1.1.1) and conditions (1.1.2), (1.1.3) we have

$$E(\xi) = \int_0^\xi \frac{\mathrm{d}E(\xi)}{\mathrm{d}\xi}\,\mathrm{d}\xi = \int_0^\xi \{\mu(l)u_z(l,\xi)u_\xi(l,\xi) - \varkappa\mu(0)[u_\xi(0,\xi)]^2\}\,\mathrm{d}\xi.$$

Taking into account the condition imposed on $u(z,\xi)$ for $z \to \infty$, for l sufficiently large when $\varkappa \geq 0$ we have $E(\xi) \leq 0$, either $E(\xi)$ is extremely small. As the integrand function in the definition of $E(\xi)$ is nonnegative, we have $E(\xi) \geq 0$. Hence it follows that $E(\xi) \equiv 0$; this yields $u(z,\xi) \equiv 0$ which was to be proved. □

Note that this proof is analogous to those given in Tikhonov and Samarskii (1966, § 1.1) for the boundary-value problem with another boundary condition.

The solution to problem 1.1 exists under the conditions of Theorem 1.2.1 formulated below. This theorem gives the explicit form of the solution of the direct problem. We suppose that

$$\frac{1}{\bar{v}^2(z)} = \frac{\eta(z)}{\mu(z)}, \qquad \bar{\sigma}(z) = \frac{\bar{v}(z)}{\mu(z)}, \qquad f(\xi) = \frac{\mathrm{d}\varphi}{\mathrm{d}\xi} \quad (\varphi(0) = 0),$$

where the function $\bar{v}(z)$ is positive and pass to the variables

$$x = \int_0^z \frac{\mathrm{d}z}{\bar{v}(z)}, \qquad \sigma(x) = \bar{\sigma}(z).$$

1.2. THE FORM OF SOLUTION OF THE DIRECT PROBLEM CONVENIENT FOR SOLVING THE INVERSE PROBLEMS

The following theorem will be used for solving the inverse problems in § 1.3, 1.4.

Theorem 1.2.1. *Denote*

$$B(x) = -\frac{1}{2}\frac{\sigma'(x)}{\sigma(x)}, \quad q(x) = \frac{3}{4}\frac{\sigma'^2}{\sigma^2} - \frac{1}{2}\frac{\sigma''}{\sigma} = B' + B^2 \tag{1.2.1}$$

(here and further the prime denotes differentiating with respect to x).

We suppose that the following conditions hold:

(A) the derivatives $\mathrm{d}^2\eta/\mathrm{d}z^2$ and $\mathrm{d}^2\mu/\mathrm{d}z^2$ are continuous for $0 \le z < \infty$.

(B) $\int_0^\infty x|q|\,\mathrm{d}x < \infty$; $\int_0^\infty B^2\,\mathrm{d}x < \infty$; $\int_0^\infty |B|\,\mathrm{d}x < \infty$ (this condition holds, for example, if for a certain finite $H > 0$ $\eta(z) \equiv \mathrm{const}$; $\mu(z) = \mathrm{const}$ for $z \ge H$; then the function $q(x)$ is finite: $q(x) \equiv 0$ for $x \ge x(H)$);

(C) the function $\varphi(\xi)$ and its derivatives up to the fourth order are continuous and absolutely integrable in the line $-\infty < \xi < \infty$.

Then the solution $u(z;\xi)$ of the direct problem 1.1 is as follows:

$$u(z,\xi)=\int_{-\infty}^{\infty}U(z,k)e^{ik\xi}\,\mathrm{d}k,\quad 0\le z<\infty,\ -\infty<\xi<\infty, \tag{1.2.2}$$

$$U(z,k)=\bar{\sigma}^{1/2}(z)W(x,k),\quad x=\int_0^z\frac{\mathrm{d}z}{\bar{v}(z)},\quad \sigma(x)=\bar{\sigma}(z), \tag{1.2.3}$$

$$W(x,k)=\frac{\gamma ik\tilde{\varphi}(k)\tilde{f}(x,k)}{\tilde{f}'(0,k)-(\alpha+ik\beta)\tilde{f}(0,k)}$$
$$(0\le x<\infty,\quad -\infty<k<\infty). \tag{1.2.4}$$

Here

$$\alpha=B(0),\quad \beta=\varkappa\bar{v}(0),\quad \gamma=\sigma^{-1/2}(0)\bar{v}(0); \tag{1.2.5}$$

the function $\tilde{f}(x,k)$ ($\operatorname{Im}k\le 0$, $x\ge 0$) solves the integral equation

$$\tilde{f}(x,k)=e^{-ikx}+\int_x^{\infty}\frac{\sin k(t-x)}{k}\,q(t)\tilde{f}(t,k)\,\mathrm{d}t; \tag{1.2.6}$$

$\tilde{\varphi}(k)$ ($-\infty<k<\infty$) is the Fourier transformation of the function $\varphi(\xi)$:

$$\varphi(\xi)=\int_{-\infty}^{\infty}\tilde{\varphi}(k)e^{ik\xi}\,\mathrm{d}k. \tag{1.2.7}$$

The integrals in formula (1.2.2) and in the equality

$$\frac{\partial u}{\partial z}=\int_{-\infty}^{\infty}\frac{\mathrm{d}U}{\mathrm{d}z}\,c^{ik\xi}\,\mathrm{d}k$$

are the Fourier integrals (i. e., the inversion formulas hold).

1.2.1. Reduction of the direct problem 1.1. We shall reduce direct problem 1.1 in the half-plane $z\ge 0$, $-\infty<\xi<\infty$ to the one-dimensional problem in the semiaxis $x\ge 0$ in the complex plane.

We shall seek for the solution $u(z,\xi)$ of direct problem 1.1 in form (1.2.2), where the integral is taken along the real axis $\operatorname{Im}k=0$ of the complex plane $k=k_1+ik_2$.

We suppose that the conditions (A) and (C) hold. Substituting (formally for now) functions (1.2.2) and (1.2.7) in the conditions of the direct problem 1.1 and changing the variables (1.2.3), we obtain the following problem for the function $W(x,k)$:

$$\frac{\mathrm{d}^2W}{\mathrm{d}x^2} + \{k^2 - q(x)\}W = 0, \qquad 0 \le x < \infty, \tag{1.2.8}$$

$$\frac{\mathrm{d}W}{\mathrm{d}x}(0,k) - (\alpha + ik\beta)W(0,k) = \gamma ik\tilde{\varphi}(k) \quad \text{for} \quad x = 0. \tag{1.2.9}$$

The function $W(x,k)$ must be analytic in the half-plane $\operatorname{Im} k < 0$, continuous up to real axis $\operatorname{Im} k = 0$, is bounded and decreases sufficiently fast for $|k| \to \infty$ in the half-plane $\operatorname{Im} k \le 0$ for each fixed value $x \in [0, \infty)$ (condition (1.2.10)).

We shall clarify the last condition. The condition of the function $W(x,k)$ being analytic with respect to k for $\operatorname{Im} k < 0$ and continuity up to the real axis $\operatorname{Im} k = 0$ is used essentially below when we establish condition (1.1.3) by means of the corollary from the Jordan lemma. The estimate for the function $W(x,k)$ for $|k| \to \infty$, which is sufficient for convergence of integral (1.2.2) and its derivatives (and sufficient for justification of all operations being done with integral (1.2.2), in particular, differentiation under the integral sum), depends on the conditions imposed on the function $\varphi(\xi)$, and will be obtained below.

Equation (1.2.8) and condition (1.2.9) are set in the real axis $-\infty < k < \infty$ of the complex plane k. The function $q(x)$ and the numbers α, β, γ are real and given by formulas (1.2.1), (1.2.5).

System (1.2.8), (1.2.9) is not corresponded to the self-adjoint Sturm—Liouville operator, since the coefficient $(\alpha + ik\beta)$ in condition (1.2.9) is complex and variable (it depends on k). (However, the differential expression $l(y) = -y'' + q(x)y$ is self-adjoint, the boundary condition is not self-adjoint).

1.2.2. Properties of the functions $f(x,k)$ and $\tilde{f}(x,k)$. By the condition (A), the function $q(x)$ is continuous for $0 \le x < \infty$. Suppose now that the first condition (B) holds:

$$\int_0^{\infty} x|q(x)|\,\mathrm{d}x < \infty. \tag{B_1}$$

If this condition holds, equation (1.2.8) has the solution $f(x,k)$ defined for $x \ge 0$, $\operatorname{Im} k \ge 0$, which satisfies the integral equation (see, Borg, 1945; Jost, 1947; Bargman, 1949; Levinson, 1949; Naymark, 1954; Faddeev, 1959)

$$f(x,k) = e^{ikx} + \int_x^{\infty} \frac{\sin k(t-x)}{k}\, q(t)f(t,k)\,\mathrm{d}t. \tag{1.2.11}$$

The solution $f(x,k)$ of equation (1.2.8) is defined by the condition

$$\lim_{x\to\infty} e^{-ikx} f(x,k) = 1.$$

Equation (1.2.8) with this condition is equivalent to integral equation (1.2.11). The analytic and asymptotic properties of the solution $f(x,k)$ are known: the function $f(x,k)$ is continuous with respect to (x,k) and is bounded for $x \geq 0$, $\operatorname{Im} k \geq 0$; for each fixed $x \in [0,\infty)$, the function $f(x,k)$ is holomorphic function with respect to k in the half-plane $\operatorname{Im} k > 0$; the function $f(x,k)$ is differentiable with respect to k up to the real axis except, may be, the point $k = 0$. We have $f(x,k) = e^{ikx} Z(x,k)$, where the function $Z(x,k)$ is bounded for $x \geq 0$, $\operatorname{Im} k \geq 0$, so that

$$|f(x,k)| < Ke^{-k_2 x}, \qquad k_2 \geq 0, \qquad x \geq 0, \qquad K = \text{const} > 0.$$

When $\operatorname{Im} k \geq 0$ and $k \to \infty$, we have

$$f_x^{(n)}(x,k) = (ik)^n e^{ikx}[1 + O(1/k)], \qquad n = 0,1,2$$

uniformly relatively $x \in [0,\infty)$. When $x \to \infty$, we have

$$f_x^{(n)}(x,k) = e^{ikx}[(ik)^n + o(1)], \qquad n = 0,1,2$$

uniformly relatively k in the half-plane $\operatorname{Im} k \geq 0$.

The function

$$\tilde{f}(x,k) = f(x,-k)$$

is defined for $x \geq 0$, $\operatorname{Im} k \leq 0$ and is the solution of equation (1.2.8). This function satisfies integral equation (1.2.6). All properties of the function $\tilde{f}(x,k)$ are obtained from the properties of the function $f(x,k)$ if we change k by $-k$. In this case, for $\operatorname{Im} k = 0$, we have

$$\tilde{f}(x,k) = \bar{f}(x,k), \qquad x \geq 0.$$

The ray $\operatorname{Im} k = 0$, $x \geq 0$ is the intersection of the domains of definitions of the functions $\tilde{f}(x,k)$ and $f(x,k)$ (if the condition (B_1) holds).

In this ray, the functions $\tilde{f}(x,k)$ and $f(x,k)$ are the fundamental system of solutions of equation (1.2.8) as the Wronskian $\mathcal{W}(f,\tilde{f}) = -2ik \not\equiv 0$.

When $\sigma(z) \equiv \text{const}$, we have $q(x) \equiv 0$ for $x \geq 0$, and the function $f(x,k)$ coincides with the solution e^{ikx} of the equation $y'' + k^2 y = 0$. The function $\tilde{f}(x,k)$ coincides with the solution e^{-ikx} of this equation. The system e^{ikx}

and e^{-ikx} is fundamental in the whole plane k; each solution is an entire function dependent on k.

We do not know anything more about the functions $f(x,k)$ and $\tilde{f}(x,k)$: if $q(x)$ is an arbitrary function, they are not expressed explicitly. However, their properties allows us to solve our problem. Note that all their properties follow from the condition (B_1) and equation (1.2.11).

1.2.3. Solution of problem (1.2.8)–(1.2.10). From condition (1.2.10) and the properties of the functions $\tilde{f}$ and f it follows that the solution $W(x,k)$ of problem (1.2.8)–(1.2.10) has the form

$$W(x,k) = D(k)\tilde{f}(x,k).$$

Determining $D(k)$ from condition (1.2.9) we obtain for $W(x,k)$ formula (1.2.4).

We show now that the function $W(x,k)$ of form (1.2.4) satisfies condition (1.2.10). In order that the function $W(x,k)$ be analytic for $\operatorname{Im} k < 0$ and continuous for $\operatorname{Im} k \le 0$, $k \ne 0$ it is necessary and sufficient that the function $A(k)$ defined for $\operatorname{Im} k \ge 0$ as follows:

$$A(k) = f'(0,k) - (\alpha - ik\beta)f(0,k) = \tilde{A}(-k),$$

where

$$\tilde{A}(k) = \tilde{f}'(0,k) - (\alpha + ik\beta)\tilde{f}(0,k) \qquad (\operatorname{Im} k \le 0)$$

has no zeros for $\operatorname{Im} k \ge 0$, $k \ne 0$.

We shall give now the scheme of the proof of this criterion.

Parallel with the function $A(k)$ defined in a certain domain D_0 of the half-plane $\operatorname{Im} k \ge 0$, we consider the set of functions

$$A_{k_0}(k) = f'(0,k) - (\alpha - ik_0\beta)f(0,k)$$

dependent on a parameter $k_0 \in D_0$. Then we establish the dependence of positions of zeros of the function $A_{k_0}(k)$ from the k_0 position (more precisely, from the sign of the imaginary part of the parameter $(\alpha - ik_0\beta)$). To this end, we connect the zeros of $A_{k_0}(k)$ with the eigenvalues of a certain (auxiliary) not self-adjoint operator from Naymark (1954) and use his theorems. This dependence is such that for each function $A_{k_0}(k)$ its zeros could not lie in D_0, i. e., the zeros could not coincide with the numbers k_0 in the parameter $(\alpha - ik_0\beta)$. As for the function $A(k)$, its zero and the number k in the coefficient $(\alpha - ik\beta)$ must coincide. Therefore, $A(k)$ has no zeros in the domain D_0. Dividing the half-plane $\operatorname{Im} k \geq 0$ into some domains (sets) D_0 and passing from one D_0 into another (the zeros $A_{k_0}(k)$ as if slipping away from the number k_0, when k_0 runs in the half-plane $\operatorname{Im} k \geq 0$), we show that the function $A(k)$ has no zeros in the whole $\operatorname{Im} k \geq 0$, $k \neq 0$.

1.2.4. The auxiliary not self-adjoint operator. Let L_{θ_0} be a not self-adjoint differential operator of the second order in the semiaxis (see, Naymark, 1954) generated by the differential expression

$$l(y) = -y'' + q(x)y, \qquad 0 \leq x < \infty$$

and the boundary condition

$$y'(0) - \theta_0 y(0) = 0.$$

Here $q(x)$ is a complex continuous function, θ_0 is a fixed complex number (independent of k). As was proved in Naymark (1954), the operator L_{θ_0} has the following interesting properties.

The sign of the imaginary part of the function $q(x)$, the sign of the imaginary part of the number θ_0, and the eigenvalues of the operator L_{θ_0} are connected as follows:

Theorem 1.2.2. *Let one of conditions*

$$\operatorname{Im} \theta_0 < 0, \ \operatorname{Im} q(x) \leq 0, \ x \geq 0 \quad \text{or} \quad \operatorname{Im} \theta_0 > 0, \ \operatorname{Im} q(x) \geq 0, \ x \geq 0$$

hold. Then all eigenvalues $\lambda = k^2$ of the operator L_{θ_0}, i.e., the squares of the zeros of the function

$$A_{\theta_0}(k) = f'(0, k) - \theta_0 f(0, k),$$

which belong to the domain $\operatorname{Im} k > 0$ may be situated only in the half-plane $\operatorname{Im} k^2 < 0$ or $\operatorname{Im} k^2 > 0$ respectively.

The disposition of zeros of the function $A_{\theta_0}(k)$ in the real axis is described by Naymark (1954).

Theorem 1.2.3. *Let one of the conditions*

$$\operatorname{Im} \theta_0 \leq 0, \ \operatorname{Im} q(x) \leq 0, \ x \geq 0 \quad \text{or} \quad \operatorname{Im} \theta_0 \geq 0, \ \operatorname{Im} q(x) \geq 0, \ x \geq 0$$

hold. Then the function $A_{\theta_0}(k)$ has no zeros in the semiaxis $k > 0$ or $k < 0$ respectively.

For not self-adjoint operator L_{θ_0}, the complex number θ_0 is fixed, and in our system (1.2.8), (1.2.9) the coefficient $(\alpha + ik\beta)$ depends on k. Therefore, we shall establish some additional statements.

1.2.5. Connection between the disposition of eigenvalues of certain sets of the operators L_{θ_0} and the lack of zeros of the function $A(k)$. We denote

$$\theta(k) = \alpha - ik\beta$$

and consider a domain D_0 from the half-plane $\operatorname{Im} k > 0$. Consider the following domain D_c. Changing the parameter k_0 in the domain D_0, we obtain the operator set $\{L_{\theta_0}\}_{D_0}$ dependent on the parameter $\theta_0 = \theta(k_0) = \alpha - ik_0\beta$, where $k_0 \in D_0$ (the function $q(x)$ for all these operators we take as in system (1.2.8)–(1.2.9)). Choose an arbitrary point k_0 from the domain D_0 and fix an operator L_{θ_0} from this set. In order to know the disposition of numbers k correspondent to the eigenvalues $\lambda = k^2$ of this operator in the half-plane $\operatorname{Im} k > 0$ we apply Theorem 1.2.2. We suppose that all these numbers (zeros of the function $A_{\theta_0}(k)$) lie in the domain $D^c_{\theta_0}$ of the half-plane $\operatorname{Im} k > 0$.

Such constructions we do for each point $k_0 \in D_0$. The union of all the domains $D^c_{\theta_0}$, where $\theta_0 = \theta(k_0)$, $k_0 \in D_0$ we denote by D_c (evidently, D_c belongs to the half-plane $\operatorname{Im} k > 0$).

Theorem 1.2.4. *If the domains D_0 and D_c do not intersect, then the function $A(k)$ has no zeros in D_0.*

Proof. Let $D_0 \cap D_c = 0$ and $A(k)$ has at most one zero $k = k_0$ in the domain D_0. This denotes that

$$l(f(x, k_0)) = k_0^2 f(x, k_0), \quad f'(0, k_0) - \theta(k_0) f(0, k_0) = 0.$$

We take an operator L_{θ_0} from the set $\{L_{\theta_0}\}_{D_0}$ assuming $\theta_0 = \theta(k_0)$. Evidently, the number k_0 corresponds to one of the eigenvalues $\lambda = k_0^2$ of the operator L_{θ_0}. Therefore, the number k_0 lies in the domain D_c which has no common points with the domain D_0. This contradiction proves the theorem. $\square$

1.2.6. The proof of the lack of zeros of the function $A(k)$ for $\operatorname{Im} k \geq 0$, $k \neq 0$. Taking into account Theorems 1.2.2 and 1.2.4 we prove that the function $A(k)$ has no zeros in the quadrants $\operatorname{Im} k > 0$, $\operatorname{Re} k > 0$ and $\operatorname{Im} k > 0$, $\operatorname{Re} k < 0$ which we take in turn as the domain D_0. In this case, the number β is nonnegative, which is essential (see formula (1.2.5)). It turns out to be that the function $\theta(k) = \alpha - ik\beta$ for $\beta \geq 0$ is such that the domains D_0 and D_c have no common points.

We take, for example, the domain $k_2 > 0$, $k_1 > 0$ as the domain D_0. Then, for $k_0 \in D_0$, we have

$$\operatorname{Im}\theta_0 = \operatorname{Im}\theta(k_0) = -k_{01}\beta < 0 \quad (\beta > 0).$$

By Theorem 1.2.2 each operator L_{θ_0} from the set $\{L_{\theta_0}\}_{D_0}$ may have eigenvalues $\lambda = k^2$ only for $\operatorname{Im}\lambda = 2k_1k_2 < 0$. In the half-plane $\operatorname{Im} k > 0$ this corresponds to the domain $k_2 > 0$, $k_1 < 0$ (the domain D_c). As $D_0 \cap D_c = 0$, then, by Theorem 1.2.4, in the domain $k_2 > 0$, $k_1 > 0$ the function $A(k)$ has no zeros. Analogously, we show that in the domain $k_2 > 0$, $k_1 < 0$ the function $A(k)$ has no zeros (in this case, the domains D_0 and D_c are changing their places).

We denote by D_0 the imaginary axis $k_1 = 0$, $k_2 > 0$. If $k_0 \in D_0$, we have $\operatorname{Im}\theta(k_0) = 0$ and obtain the set of self-adjoint operators $\{L_{\theta_0}\}_{D_0}$. Therefore, the squares of zeros of the function $A(k)$, for $k \in D_0$ may be only real and negative, as, for $k \in D_0$, we have $k^2 = -k_2^2 < 0$. However, each operator L_{θ_0} from the set $\{L_{\theta_0}\}_{D_0}$ is positive definite for $\beta \geq 0$ (this can be established following A. S. Alekseev); therefore, L_{θ_0} has no negative values $\lambda = k^2$. Positive definiteness of each operator L_{θ_0} from the set $\{L_{\theta_0}\}_{D_0}$ when $\beta \geq 0$ we establish analogously Alekseev (1962).

The operator L_{θ_0} is generated by the differential expression

$$l(y) = -\frac{\mathrm{d}^2}{\mathrm{d}x^2}\, y + q(x)y,$$

where

$$q(x) = \frac{\mathrm{d}B}{\mathrm{d}x}(x) + B^2(x),$$

and by the boundary condition

$$y'(0) - \theta_0 y(0) = 0.$$

Here $\theta_0 = \theta(k_0) = \alpha + \beta k_{02}$ (when $k_0 \in D_0$, $k_0 = (k_{01} + ik_{02})$, we have $k_{01} = 0$, $k_0 = ik_{02}$), $\alpha = B(0)$. Therefore, we have

$$(L_{\theta_0}y, y) = \int_0^\infty \Big\{\Big[-\frac{\mathrm{d}}{\mathrm{d}x^2} + q(x)\Big]y\Big\}\bar{y}\,\mathrm{d}x$$
$$= \int_0^\infty \Big\{\Big[\frac{\mathrm{d}}{\mathrm{d}x} + B(x)\Big]\Big[-\frac{\mathrm{d}}{\mathrm{d}x} + B(x)y\Big]\Big\}\bar{y}\,\mathrm{d}x$$
$$= \int_0^\infty \Big[-\frac{\mathrm{d}}{\mathrm{d}x} + B(x)\Big]y \cdot \overline{\Big[-\frac{\mathrm{d}}{\mathrm{d}x} + B(x)\Big]y}\,\mathrm{d}x + \Big\{\Big[-\frac{\mathrm{d}}{\mathrm{d}x} + B(x)\Big]y\Big\}\bar{y}\,\Big|_{x=0}^{\infty}.$$

Here the first term is nonnegative. The second term is nonnegative also, since, by the boundary condition, we have

$$\Big\{-\frac{\mathrm{d}y}{\mathrm{d}x}+B(x)y\Big\}\Big|_{x=0}=\beta k_{02}y(0).$$

Hence we obtain the inequality $(L_{\theta_0}y,y)\geq 0$. In this case we have set

$$\Big\{\Big[-\frac{\mathrm{d}y}{\mathrm{d}x}+B(x)\Big]y\Big\}\bar{y}\Big|_{x=\infty}=0.$$

The function $y=f(x,k)$ and its derivative $f'(x,k)$ decreases as e^{-k_2x} (see point *b*) when $\operatorname{Im}k=k_2>0$ for $x\to\infty$; therefore, it satisfies this condition.

Hence it follows that the function $A(k)$ has no zeros for $k_1=0$, $k_2>0$, $\beta\geq 0$.

When we establish the lack of zeros of $A(k)$ in the real axis $\operatorname{Im}k=0$, $k\neq 0$, instead of Theorem 1.2.2 we apply Theorem 1.2.3. We consider now a set M_0 in the axis $\operatorname{Im}k=0$ and the set of functions $\{A_{\theta_0}(k)\}_{M_0}$ dependent on the parameter $\theta_0=\theta(k_0)$, where $k_0\in M_0$. Assume that each function $A_{\theta_0}(k)$ from this set has no zeros in a set M_n from the axis $\operatorname{Im}k=0$. The role of Theorem 1.2.4 now plays

Theorem 1.2.5. *If $M_0=M_n$ or $M_n\subset M_0$, then the function $A(k)$ has no zeros in the set M_n.*

We take now the semiaxis $k>0$ as the set M_0. Then, for $k_0\in M_0$, we have $\operatorname{Im}\theta_0=-k_0\beta\leq 0$ for $\beta\geq 0$. By Theorem 1.2.3, each function $A_{\theta_0}(k)$ from the set $\{A_{\theta_0}(k)\}_{M_0}$ has no zeros in the semiaxis $k>0$ (the set M_n). Therefore, $M_0=M_n$ and, by Theorem 1.2.5, the function $A(k)$ has no zeros for $k>0$. Analogously we can show that $A(k)$ has no zeros in the semiaxis $k<0$.

1.2.7. The point $k=0$. Special attention should be given to the point $k=0$. The following statement holds.

Theorem 1.2.6. *Suppose conditions (A) and (B) of Theorem 1.2.1 hold. Then the numbers*

$$z_1=\lim_{k\to 0}-(f'(0,k)+\alpha f(0,k))/(ik)\quad\text{and}\quad y_1=f(0,0)$$

are finite, real; they are not equal to zero and have various signs.

Corollary 1.2.1. *The limit*

$$\lim_{k\to 0}\{-A(k)/(ik)\} = z_1 - \beta y_1$$

exists, is finite, is not equal to zero for $\beta \geq 0$. So, the function $W(x,k)$ is continuous in the point $k = 0$ for $x \geq 0$.

Proof. The proof of Theorem 1.2.6 uses Lemma 2.1.1 (see Naymark, 1954). We give its formulation.

Lemma 1.2.1. *Consider the integral equation*

$$u(x,s) = F(x,s) + \int_a^\infty K(x,\xi,s)u(\xi,s)\,\mathrm{d}\xi, \qquad (*)$$

whose kernel $K(x,\xi,s)$ satisfies the following conditions.

a) *For each value s from a certain set E of the complex s-plane and for each value x from the interval $[a,\infty)$, the function $K(x,\xi,s)$ is a measurable integrable function of the variable ξ in the interval $[a,\infty)$.*

b) *There exists a positive number $g < 1$ so that*

$$\int_a^\infty |K(x,\xi,s)|\,\mathrm{d}\xi < g$$

for all $x \in [a,\infty)$ and $s \in E$.

c) *For each point $x_0 \in [a,\infty)$ and $s_0 \in E$, we have*

$$\lim_{\substack{x\to x_0\\ s\to s_0}} \int_a^\infty |K(x,\xi,s) - K(x_0,\xi,s_0)|\,\mathrm{d}\xi = 0,$$

$$a \leq x, \qquad x_0 < \infty, \qquad s, s_0 \in E.$$

The condition a), b), c) will be a fortiori hold if

$$K(x,\xi,s) = \begin{cases} \varphi_1(x,\xi,s)p(\xi) & \text{for} \quad a \leq \xi \leq x,\\ \varphi_2(x,\xi,s)p(\xi) & \text{for} \quad \xi > x \geq a.\end{cases}$$

Here $p(\xi)$ is integrable function in the interval $[a,\infty)$; the functions $\varphi_1(x,\xi,s)$, $\varphi_2(x,\xi,s)$ are bounded continuous functions with respect *to the variables (x, ξ, s), $a \le x$, $\xi < \infty$, $s \in E$. These functions satisfy the conditions*

$$c \int_a^\infty |p(\xi)| \, d\xi < 1$$

for

$$c = \max\{\sup |\varphi_1(x, \xi, s)|, \sup |\varphi_2(x, \xi, s)|\}.$$

Assume also that:

d) The function $F(x, s)$ is continuous and bounded for $a \le x < \infty$, $s \in E$.

If the conditions a), b), c), d) hold then the equation (*) has the solution $u(x, s)$ continuous with respect to (x, s) and bounded for $a \le x \le \infty$, $s \in E$.

From integral equation (1.2.11) and the properties of the function $f(x, k)$ (see Subsection 1.2.2) it follows that the function

$$f_1(x, k) = (-f'(x, k) + B(x) f(x, k))/(ik)$$

satisfies the integral equation

$$f_1(x, k) = \Big\{-e^{ikx} - \int_x^\infty B(t) \sin k(t - x) f(t, k) \, dt\Big\} + \int_x^\infty B(t) \cos k(t - x) f_1(t, k) \, dt.$$

The function from figure brackets playing the role of free term multiplied by e^{-ikx} is continuous with respect to (x, k) and bounded for $x \in [0, \infty)$, $\operatorname{Im} k \ge 0$. This follows from the condition (B) and from boundedness of the function $\{e^{-ikx} f(x, k)\}$. From Lemma 1.2.1 it follows that the function $z_1(x, k) = e^{-ikx} f_1(x, k)$ and the function $f_1(x, k)$ are bounded and continuous with respect to (x, k) for $x \ge 0$, $\operatorname{Im} k \ge 0$.

Evidently, the function

$$f_1(x) = f_1(x, 0) = \lim_{k \to 0} f_1(x, k)$$

satisfies the integral equation

$$f_1(x) = -1 + \int_x^\infty B(t) f_1(t) \, dt, \quad x \ge 0.$$

Hence it follows that $f_1(x) \not\equiv 0$ and

$$-f_1(x) + f_1(0) = \int_0^x B(t) f_1(t) \, dt.$$

The last equation is the integral equation of Volterra type and the function $f_1(x)$ is its bounded solution. Therefore, the number $z_1 = f_1(0) \neq 0$. Analogously, for the function $y_1(x) = f(x, 0)$, we have

$$y_1(x) = 1 - \int_x^\infty B(t) y_1(t)\,dt \qquad \text{and} \qquad y_1(0) \neq 0.$$

The identity

$$\int_0^x f_1(x,k)\bar{f}^*(x,k)\,dx + f_1(x,k)\bar{f}(x,k) - f_1(0,k)\bar{f}(0,k) \\ = -ik\int_0^x f(x,k)\bar{f}(x,k)\,dk,$$

where

$$f^*(x,k) = -f'(x,k) + B(x)f(x,k)$$

and the properties of the functions $f(x,k)$, $f_1(x,k)$ yield for $k \to 0$:

$$f_1(x)y_1(x) - f_1(0)y_1(0) = 0 \qquad 0 \le x \le a < \infty.$$

For $x \to \infty$ we have $y_1(x) \to 1$, $f_1(x) \to (-1)$.

Therefore, the last equality for a and x sufficiently large gives that $z_1 = f_1(0)$ and $y_1 = y_1(0)$ have various signs. Theorem 1.2.6 is proved. □

1.2.8. Completion of the proof of Theorem 1.2.1 and properties of the solution $u(z,\xi)$. Properties of the function $\tilde{f}(x,k)$ (see Subsection 1.2.2) and formula (1.2.4) yield the relations

$$|W(x,k)| < M|\tilde{\varphi}(k)|, \qquad \operatorname{Im} k \le 0, \quad x \ge 0; \tag{1.2.12}$$

$$W(x,-k) = \bar{W}(x,k), \qquad -\infty < k < \infty, \quad x \ge 0. \tag{1.2.13}$$

We suppose that the function $\varphi(\xi)$ and its derivatives up to the second order are continuous and absolutely integrable in the line $-\infty < \xi < \infty$. The last condition for the function $\varphi(\xi)$ denotes that

$$\int_{-\infty}^{\infty} |\varphi(\xi)|\,d\xi < \infty.$$

The set (the class) of all absolutely integrable functions in the line $-\infty < \xi < \infty$ is denoted by $L_1(-\infty;\infty)$.

It is known (see, for example, Shilov, 1970, pg. 277) that there exists the connection between the smoothness of the function $\varphi(\xi)$ and the speed of decreasing of its Fourier transformation $\tilde{\varphi}(k)$ when $|k| \to \infty$.

We denote by $F[\varphi]$ the Fourier transformation of the function $\varphi(\xi)$:

$$F[\varphi] = \int_{-\infty}^{\infty} \varphi(\xi) e^{-ik\xi} \, d\xi.$$

We suppose that the function $\varphi(\xi)$ and its derivatives up to the order n are continuous and absolutely integrable in the axis $-\infty < \xi < \infty$. Then, for the Fourier transformation of derivatives $\varphi^{(m)}(\xi)$, the equality

$$F[\varphi^{(m)}(\xi)] = (ik)^m F[\varphi], \quad m = 0, 1, \dots, n$$

and the estimate

$$|F[\varphi]| = |F[\varphi^{(m)}(\xi)]|/|k|^m \leq c/|k|^m$$

hold.

Taking into account inequality (1.2.2), we obtain the estimate for $|k| \to \infty$

$$|W(x,k)| < M/|k|^2, \qquad x \geq 0.$$

Therefore, the function $W(x,k)$ belongs to the class $L_1(-\infty, \infty)$ for each fixed $x \in [0, \infty)$. Hence it follows that the complex function

$$U(z,k) = \sigma^{1/2}(z) W(x,k)$$

dependent on real variable $k \in (-\infty, \infty)$ is represented by the Fourier integral (By Theorem 23, see, E. K. Titchmarsh, 1948). This denotes that the integral in (1.2.2) for the function $u(z,\xi)$, which is the Fourier transformation of the complex function $U(z,k)$, has meaning and is the Fourier integral. Relation (1.2.13) yields that $u(z,\xi)$ is a real function.

The function $u(z,\xi)$ defined by formula (1.2.2) satisfies initial condition (1.1.3) since the function $U(z,k)$ defined by equalities (1.2.3)–(1.2.7) satisfies the condition of the theorem formulated below. This theorem is the corollary of the Jordan lemma well-known in complex function theory.

Theorem 1.2.7. *Let a function $f(k)$ be analytic in the half-plane* $\operatorname{Im} k < 0$, *be continuous up to the real axis $-\infty < k < \infty$, and tend to zero when $|k| \to \infty$ uniformly relatively* $\arg k$ $(\pi \leq \arg k \leq 2\pi)$. *Then the integral*

$$\int_{-\infty}^{\infty} f(k) e^{ik\xi} \, dk$$

is equal to zero for each $\xi < 0$.

We consider now properties of the function $u(z,\xi)$ of form (1.2.2), where $U(z,k)$ is defined by formulas (1.2.3), (1.2.4). Evidently, we have

(a) $U(z,k) \in L_1(-\infty,\infty)$ for each $x \in [0,\infty)$;

(b) integral (1.2.2) converges uniformly relatively $x \in [0,\infty)$ and $\xi \in (-\infty,\infty)$.

The property (a) yields

(c) if $\bar{v}(z) < M$ for $z \geq 0$, then the function $u(z,\xi)$ is uniformly bounded for all $z \in [0,\infty)$, $\xi \in (-\infty,\infty)$.

The property (a), the Riemann—Lebesque theorem, and the formula

$$W(x,k) = \frac{\gamma ik\tilde{\varphi}(k)\tilde{Z}(x,k)}{\tilde{A}(k)}\,e^{-ikx},$$

where $\tilde{Z}(x,k) = Z(x,-k)$ and $|\tilde{Z}(x,k)| < M$, yield

(d) function (1.2.2) tends to zero uniformly relatively $z \in [0,\infty)$ or $\xi \in (-\infty,\infty)$ when $\xi \to \pm\infty$ or $z \to \infty$ respectively.

Taking into account the properties of integral dependence on the parameter we obtain the property

(e) function (1.2.2) is continuous with respect to (z,ξ) in the half-plane $0 \leq z < \infty$, $-\infty < \xi < \infty$.

When we have deduced system (1.2.8)–(1.2.10), we have used the relations

$$\frac{\partial^n u}{\partial z^n}(z,\xi) = \int_{-\infty}^{\infty} \frac{\mathrm{d}^n U}{\mathrm{d}z^n}(z,k)e^{ik\xi}\,\mathrm{d}k, \quad n = 1,2; \tag{1.2.14}$$

$$\frac{\partial^n u}{\partial \xi^n}(z,\xi) = \int_{-\infty}^{\infty} (ik)^n U(z,k)e^{ik\xi}\,\mathrm{d}k, \quad n = 1,2. \tag{1.2.15}$$

Using the estimates

$$\left|\frac{\mathrm{d}^n U}{\mathrm{d}z^n}(z,k)\right| < M|k|^n|\tilde{\varphi}(k)|, \qquad |(ik)^n U(z,k)| < M|k|^n|\tilde{\varphi}(k)|,$$

and the connection between smoothness of $\varphi(\xi)$ and decreasing $\tilde{\varphi}(k)$ we establish the formulas of integrand differentiating (1.2.14), (1.2.15). Similarly, as for the function $u(z,\xi)$, we can show that each partial derivative of the function $u(z,\xi)$ in (1.2.14), (1.2.15) satisfy properties (a)–(e).

So, using the properties of the function $\tilde{f}(x,k)$ and the properties of the integral dependent on the parameter we have shown that the function $u(z,\xi)$ has the properties which are required in the direct problem 1.1: the smoothness of $u(z,\xi)$ and behavior in the infinity. Thus, we have established that the function $u(z,\xi)$ defined by formulas (1.2.2)–(1.2.7) is a real solution of direct problem 1.1. Theorem 1.2.1 is thus proved.

Simultaneously we have shown the existence of solution of direct problem 1.1.

1.3. THE INVERSE PROBLEM WITH THE DATA $u(0,\xi)$ AND $\partial u/\partial z|_{z=0}$

Inverse problem 1.1. Let the values $u(0,\xi)$ and $\partial u/\partial z|_{z=0}$ of the function $u(z,\xi)$ be given for $0 \le \xi < \infty$. It is known that $u(z,\xi)$ is a solution of the direct problem 1.1 with certain unknown functions $\eta(z)$, $\mu(z)$ $(0 \le z < \infty)$, $f(\xi)$ $(0 < \xi < \infty)$ and unknown number $\varkappa \ge 0$. We know also the value $\mu(0)$.

We need to find the function $\sigma(x)$ for $0 \le x < \infty$ (in the general case); if $\mu(z) \equiv \mu(0)$ then we seek for the function $\eta(z)$ for $0 \le z < \infty$; if the function $\bar{v}(z) = \mu(z)/\eta(z)$ is known for $0 \le z < \infty$, then we must find the functions $\eta(z)$ and $\mu(z)$ for $0 \le z < \infty$.

Solution and the uniqueness theorem for inverse problem 1.1. Parallel with our not self-adjoint differential system (1.2.8), (1.2.9) we consider the following self-adjoint Sturm—Liouville system:

$$\frac{d^2y}{dx^2} + \{k^2 - q(x)\}y = 0, \qquad 0 \le x < \infty, \tag{1.3.1}$$

$$y(0,k) = 1, \qquad y'(0,k) = \alpha. \tag{1.3.2}$$

Here the function $q(x)$ and the number α are taken from system (1.2.8), (1.2.9). Further, we consider the relation

$$\begin{aligned} d\rho(\lambda) &= -\frac{i}{\pi}\Big\{\frac{k\tilde{f}(0,k)}{\tilde{f}'(0,k) - \alpha\tilde{f}(0,k)} - \frac{kf(0,k)}{f'(0,k) - \alpha f(0,k)}\Big\}\,dk \\ &= \frac{2k^2\,dk}{\pi|f'(0,k) - \alpha f(0,k)|^2} \end{aligned} \tag{1.3.3}$$

in the real axis $-\infty < k < \infty$ ($\lambda = k^2 \geq 0$). As is known (see Alekseev, 1962), $\rho(\lambda)$ ($\lambda \geq 0$; $\rho(0) = 0$) is the spectral function of system (1.3.1), (1.3.2).

Our inverse problem 1.1 we reduce to the Sturm—Liouville inverse problem for system (1.3.1), (1.3.2). The spectral function $\rho(\lambda)$ of this system we define using the data of our inverse problem. For this purpose we apply Theorem 1.2.1.

Firstly, inverting the Fourier integrals in the formulas

$$u(0,\xi) = \int_{-\infty}^{\infty} U(0,k)e^{ik\xi}\,\mathrm{d}k, \qquad \frac{\partial u}{\partial z}\Big|_{z=0} = \int_{-\infty}^{\infty} \frac{\mathrm{d}U}{\mathrm{d}z}\Big|_{z=0} e^{ik\xi}\,\mathrm{d}k$$

(see Theorem 1.2.1), we find the functions

$$U(0,k) = \sigma^{1/2}(0)W(0,k), \tag{1.3.4}$$

$$\frac{\mathrm{d}U(z,k)}{\mathrm{d}z}\Big|_{z=0} = \frac{\sigma^{1/2}(0)}{\bar{v}(0)}\Big[\frac{\mathrm{d}W}{\mathrm{d}x}(0,k) - \alpha W(0,k)\Big] \tag{1.3.5}$$

uniquely in the whole real axis $-\infty < k < \infty$ of the plane k. Taking into account formula (1.2.4), we have the relations

$$kW(0,k) = \frac{k\tilde{f}(0,k)\gamma ik\tilde{\varphi}(k)}{\tilde{f}'(0,k) - (\alpha + ik\beta)\tilde{f}(0,k)}, \tag{1.3.6}$$

$$\frac{\mathrm{d}W}{\mathrm{d}x}(0,k) - \alpha W(0,k) = \frac{\{\tilde{f}'(0,k) - \alpha\tilde{f}(0,k)\}\gamma ik\tilde{\varphi}(k)}{\tilde{f}'(0,k) - (\alpha + ik\beta)\tilde{f}(0,k)}. \tag{1.3.7}$$

From the results of § 1.2 it follows that the function

$$A_0(k) = A(k)|_{\beta=0} = f'(0,k) - \alpha f(0,k)$$

does not vanish for $\operatorname{Im} k \geq 0$, $k \neq 0$, and the limit $lim_{k\to 0}(A_0(k))/k$ is not equal to zero and is finite. Using this fact and formulas (1.3.6), (1.3.7), and dividing equality (1.3.4) by equality (1.3.5), we obtain the function

$$\frac{kU(0,k)}{\mathrm{d}U(z,k)/\mathrm{d}z|_{z=0}} = \bar{v}(0)\,\frac{k\tilde{f}(0,k)}{\tilde{f}'(0,k) - \alpha\tilde{f}(0,k)} = \tilde{F}(k) \tag{1.3.8}$$

for all real values $k \in (-\infty,\infty)$. Then we find the function (see formula (1.3.3)):

$$-(i/\pi)\{\tilde{F}(k) + \tilde{F}(-k)\}\,\mathrm{d}k = \bar{v}(0)\,\mathrm{d}\rho(\lambda) \tag{1.3.9}$$

for $-\infty < k < \infty$, $\lambda \geq 0$. After, we define the value $\bar{v}(0)$ calculating the integral

$$G(t) = \bar{v}(0) \int_0^\infty \frac{\sin kt}{k} \, d\rho(\lambda) = \bar{v}(0)\Phi'(t)$$

for $t \in [0, \varepsilon)$ $(\varepsilon > 0)$ and applying the property (see Krein, 1963; Alekseev, 1962) $\Phi'(0) = \lim_{t\to 0} \Phi'(t) = 1$ of the function $\Phi'(t)$. The prime here denotes differentiation with respect to t. Using the value $\bar{v}(0)$, from expression (1.3.9) we derive the spectral function $\rho(\lambda)$, $\lambda \geq 0$ of system (1.3.1), (1.3.2) or, for $t \geq 0$, the function

$$\Phi(t) = \int_0^\infty \frac{1 - \cos\sqrt{\lambda}t}{\lambda} \, d\rho(\lambda).$$

This function is the main transition function (Krein, 1953, 1954) of this system. Using the methods of Gel'fand and Levitan (1951), Marchenko (1952), Krein (1954) for solution of inverse Sturm—Liouville problem, we find the function $q(x)$ for $0 \leq x < \infty$ and the number α by the function $\rho(\lambda)$ or $\Phi(t)$.

Given the number $\mu(0)$ and $\bar{v}(0)$ which we had found, we obtain $\sigma(0) = \bar{v}(0)/\mu(0)$. Finally, we find the function $\sigma(x)$ for $0 \leq x < \infty$, applying, for example, the explicit formula $\varphi(x) = \sigma^{1/2}(0) \cdot \sigma^{-1/2}(x)$, where $\varphi(x)$ is the solution of equation

$$-y'' + q(x)y = 0$$

with the initial data

$$y(0) = 1; \qquad y'(0) = \alpha.$$

If, besides, $\mu(z) \equiv \mu(0)$, then $\mu(0)\sigma(x) = \tilde{v}(x)$, and we can reestablish subsequently the functions

$$z(x) = \int_0^x \tilde{v}(x) \, dx \quad \text{and} \quad \bar{v}(z) = \tilde{v}(x),$$

where $z = z(x)$; therefore, we find the function $\eta(z)$ for $0 \leq z < \infty$. If we know the function $\bar{v}^2(z) = \mu(z)/\eta(z)$ for $0 \leq z < \infty$, then we find the functions

$$x = \int_0^z \frac{dz}{\bar{v}(z)}, \qquad \tilde{v}(x) = \bar{v}(z),$$

where

$$x = x(z), \quad \mu(x) = \tilde{v}(x)/\sigma(x) \quad \text{for} \quad 0 \leq x < \infty,$$

and recover the functions $\mu(z)$ and $\eta(z)$ for $0 \leq z < \infty$.

Note that the spectral function $\rho(\lambda)$ (and the main transition function $\Phi(t)$ also) of differential system (1.3.1), (1.3.2) is determined uniquely from the data of our inverse problem which follows from the method of its solution. By the Marchenko theorem (see, Marchenko, 1952), given the spectral function $\rho(\lambda)$, the function $q(x)$ for $0 \leq x < \infty$ and the number α are defined uniquely also. Following the results of Krein (1954), these values are obtained uniquely by the given function $\Phi(t)$. Therefore, the following theorem holds.

Theorem 1.3.1. *Let the conditions (A), (B), (C) of Theorem 1.2.1 hold. Then the inverse problem 1.1 has a unique solution for each* $\varkappa \geq 0$.

1.4. THE INVERSE PROBLEM FOR SEMIBOUNDED STRING WHICH HAS NO ANALOG IN THE CASE $\varkappa = 0$

The results of this section will be used in Chapter 3.

The inverse problem 1.1 can be considered both for the case $\varkappa > 0$ (the boundary condition for the directional derivative) and for the case $\varkappa = 0$ (the boundary condition for the normal derivative). Unlike this, the inverse problem considered below holds only for $\varkappa \neq 0$ and has no analog for the case $\varkappa = 0$. Besides, in this inverse problem, we should not set the number $\mu(0)$ (which is important for applications considered in Chapter 3).

We introduce now the number $\delta > 0$ and the function $\varphi^*(\xi)$ as follows:

$$\varkappa = \delta/\mu(0), \qquad \varphi(\xi) = \varkappa\varphi^*(\xi), \qquad \varkappa > 0. \tag{1.4.1}$$

Inverse problem 1.2. We suppose that the function $\varphi^*(\xi)$ and the values $u(0,\xi)$ of a solution $u(z,\xi)$ of the direct problem 1.1 with unknown functions $\eta(z)$, $\mu(z)$ $(0 \leq z < \infty)$ and unknown numbers $\varkappa > 0$ are set for $0 \leq \xi < \infty$. The number $\delta > 0$ is known. (The function $\partial u/\partial z|_{z=0}$ set in the inverse problem 1.1, evidently, is not determined from condition (1.1.2) by these data).

We need to find in the general case — the function $\sigma(x)$ when $0 \leq x < \infty$. If $\mu(z) \equiv \text{const} = \mu$ and the number μ is known, then we need to find the function $\eta(z)$ for $0 \leq z < \infty$. If the function $\bar{v}(z)$ is known for $0 \leq z < \infty$, then we must find the function $\eta(z)$ and $\mu(z)$ for $0 \leq z < \infty$.

Taking into account relations (1.4.1), we obtain from Theorem 1.2.1:

$$U(z,k) = \frac{\sigma^{1/2}(z)\gamma ik\tilde{\varphi}^*\tilde{f}(x,k)}{\tilde{f}'(0,k) - (\alpha + ik\beta)\tilde{f}(0,k)},$$

where

$$\alpha = -\frac{1}{2}\frac{\sigma'(0)}{\sigma(0)}, \qquad \beta = \delta\sigma(0), \qquad \gamma = \beta\sigma^{-1/2}(0),$$

and $\tilde{\varphi}^*(k)$ is the Fourier transformation of the function $\varphi^*(\xi)$.

Similar to the case of the inverse problem 1.1, we reduce this problem to the inverse Sturm—Liouville problem. However, these are some distinctions from the methods of § 1.3. Namely, in order to find the function $\rho(\lambda)$, we apply the formula

$$\frac{kU(0,k)}{U(0,k) + \tilde{\varphi}(k)} = \frac{k\tilde{f}(0,k)}{\tilde{f}'(0,k) - \alpha\tilde{(}0,k)}$$

instead of (1.3.8).

Then, analogously § 1.3, given $\rho(\lambda)$, we find the function $q(x)$ for $0 \le x < \infty$ and the number α. The number $\sigma(0)$ is determined otherwise also. First, using the values $q(x)$ and α obtained before, we construct the function $\tilde{f}(x,k)$ and find the number β from the equality

$$U(0,k) = \frac{\beta ik\tilde{\varphi}^*(k)\tilde{f}(0,k)}{\{\tilde{f}'(0,k) - \alpha\tilde{f}(0,k)\} - ik\beta\tilde{f}(0,k)};$$

then we find the number $\sigma(0)$ from the formula $\sigma(0) = \beta/\delta$. At last, the solution of the inverse problem 1.2 is obtained following the methods of § 1.3, and the theorem below holds.

Theorem 1.4.1. *Let the conditions of Theorem 1.2.1 hold. Then, the inverse problem 1.2 has a unique solution for each* $\delta > 0$.

Chapter 2.

Inverse problems for the elliptic equation in the half-plane

We shall consider here the coefficient inverse problem for the elliptic equation

$$\frac{\partial^2 u}{\partial z^2} + \eta(z)\,\frac{\partial^2 u}{\partial \xi^2} + \tau(z)\,\frac{\partial u}{\partial z} = 0, \tag{2.1}$$

set in the half-plane $z \geq 0$, $-\infty < \xi < \infty$ given certain data on the solution of the boundary-value problem (direct) for this equation with the directional derivative condition

$$\left(\frac{\partial u}{\partial z} - \varkappa\frac{\partial u}{\partial \xi}\right)\Big|_{z=0} = f(\xi) \qquad (-\infty < \varkappa < \infty)$$

set on the boundary $z = 0$. The data is given similarly as in the inverse problems considered above for the hyperbolic equation for string vibration. It is

either the solution of the boundary-value problem and its normal derivative given for the points of the boundary;

or the solution of the boundary-value problem given for the boundary points and the function $f(\xi)/\varkappa$.

Both inverse problems are reduced to the known inverse Sturm—Liouville problem: to find the self-adjoint Sturm—Liouville operator by (1) the principal transition function or by (2) the spectral function in spite of the fact that the correspondent one-dimensional operator is not self-adjoint and is not Sturm—Liouville. (In the first case, thus, the inverse problems for elliptic equation are reduced to the known inverse problem for the hyperbolic equation of string vibration). This allows us to formulate and prove the uniqueness theorem for the inverse problems in question and suggest the method to solve them.

The similar approach was applied in the first chapter for the hyperbolic equation and generates the analogy in conditions of the uniqueness theorems for elliptic equation (2.1) and the inverse problems of Chapter 1 for the string. However, this proof contains a new element (in comparison with Chapter 1): when we determine the principal transition or spectral function, we have to use the method of analytic continuation.

2.1. FORMULATION OF THE DIRECT PROBLEM

Boundary-value (direct) problem 2.1. Let the equation

$$\frac{\partial^2 u}{\partial z^2} + \eta(z)\,\frac{\partial^2 u}{\partial \xi^2} + \tau(z)\,\frac{\partial u}{\partial z} = 0 \quad (0 < m < \eta(z) < M < \infty) \tag{2.1.1}$$

be given in the domain $z > 0$, $-\infty < \xi < \infty$, and the boundary condition

$$\left(\frac{\partial u}{\partial z} - \varkappa\frac{\partial u}{\partial \xi}\right)\Big|_{z=0} = f(\xi) \quad (-\infty < \varkappa < \infty) \tag{2.1.2}$$

be given for $z = 0$, $-\infty < \xi < \infty$.

We suppose that the following conditions hold also:

the function $u(z,\xi)$ and the derivatives u_z, u_ξ are continuous in the half-plane $z \geq 0$, $-\infty < \xi < \infty$; and the second order derivatives are continuous in the domain $z > 0$, $-\infty < \xi < \infty$;

the function $u(z,\xi)$ tends to zero uniformly relatively $\xi \in (-\infty,\infty)$ and $z \in [0,\infty)$ respectively when $z \to \infty$ or $|\xi| \to \infty$;

the coefficients of equation (2.1.1) are bounded for $0 < z < \infty$.

We need to find the function $u(z,\xi)$ defined in the half-plane $z \geq 0$, $-\infty < \xi < \infty$ satisfying all the above conditions.

Such problem we shall call the boundary-value problem or the direct problem 2.1. Unlike the inverse problems considered below (§ 2.3) the number $\varkappa$ $(-\infty < \varkappa < \infty)$ and the functions $\eta(z)$, $\tau(z)$ $(0 \le z < \infty)$, $f(\xi)$ $(-\infty < \xi < \infty)$ are given in the direct problem.

The solution of the boundary-value problem 2.1 exists under the conditions of Theorem 2.2.1 formulated below. The following theorem holds.

Theorem. *The boundary value problem 2.1 may have only one solution.*

Proof. The formulation of the boundary-value problem 2.1 yields that its solution $u(z,\xi)$ satisfies the following condition (*):

> the function $u(z,\xi)$ tends to zero when $r = \sqrt{z^2+\xi^2} \to \infty$ uniformly relatively the points of the semicircle of the radius r with the center in the point $z = \xi = 0$.

The boundary condition (2.1.2) we shall write as follows:

$$(\vec{l} \cdot \operatorname{grad} u(z,\xi)) = \frac{\partial u}{\partial l} = f(\xi), \quad z = 0, \quad -\infty < \xi < \infty,$$

where the vector $\vec{l} = \vec{i}_z - \varkappa \vec{i}_\xi$ is given in the boundary $z = 0$ of the domain $z \ge 0$, $-\infty < \xi < \infty$ with the extremal normal $\vec{N} = -1 \cdot \vec{i}_z$. So, everywhere in the boundary $z = 0$ we have

$$(\vec{l} \cdot \vec{N}) = -1 \ne 0, \qquad -\infty < \xi < \infty.$$

Here we have denoted $\vec{i}_z$, $\vec{i}_\xi$ the basis vectors of the axes z and ξ.

We suppose that the two solutions $u_1(z,\xi)$ and $u_2(z,\xi)$ of the boundary-value problem 2.1 exist. The function $w(z,\xi) = u_1(z,\xi) - u_2(z,\xi)$ satisfies equation (2.1.1), the condition (*) and the condition:

$$\frac{\partial w}{\partial l} = (\vec{l} \cdot \operatorname{grad} w) = 0, \quad -\infty < \xi < \infty. \qquad (**)$$

If the function w is constant in the half-plane $z = 0$, then the condition (*) yields $w(z,\xi) = 0$.

Suppose that in the half-plane $z \ge 0$ the function w is not equal to a constant identically. We consider in this half-plane a certain domain D with a smooth boundary S (without the break points) consisted from two parts. The part S_0 is the segment $-a \le \xi \le a$ $(a > 0)$ of the boundary $z = 0$, and the remained part $S_1 = S \setminus S_0$. We denote by R the diameter of the domain D: $R = \min_{(z,\xi)\in S_1} \sqrt{z^2+\xi^2}$.

By the condition (*), for each $\varepsilon > 0$, there exists a number $R_\varepsilon > 0$, so that

$$|w(z,\xi)| < \varepsilon$$

everywhere in the part S_1 of the boundary S of each domain $D = D_R$ with the diameter $R > R_\varepsilon$. We shall show that this inequality holds also in the part S_0 of the boundary S of the domain D_R.

As in the domain D_R: $w(z,\xi) \not\equiv \text{const}$, then, by the Hopf principle for elliptic equations (see Bitsadze, 1966; Hopf, 1927; Miranda, 1957), we have $w(z,\xi) \neq \text{const}$ in the boundary S of this domain D_R. Therefore, the function w has the minimum point ($w_{\min} = m_0$) and the maximum point ($w_{\max} = M_0$) in the boundary S. We suggest that at least one of this extremal points is situated in the part S_0. By the Zaremba—Giraud principle for elliptic equations (see Bitsadze, 1966; Zaremba, 1927; Giraud, 1932; Hopf, 1952; Oleinik, 1952), $\partial w/\partial l > 0$ along each nontangential directions $\vec{l}$ in the minimum point and $\partial w/\partial l < 0$ in the maximum point. This contradicts the identity (**) in the boundary $z = 0$. Therefore, both this points belong to the part S_1 of the boundary S, where the condition $|w(z,\xi)| < \varepsilon$ holds.

Hence it follows that for all the points of S_0 we have

$$-\varepsilon < m_0 < w(z,\xi) < M_0 < \varepsilon,$$

i. e., everywhere in the boundary $S = S_0 \cup S_1$ of the domain D_R the inequality $|w| < \varepsilon$ holds. By the Hopf principle, this inequality holds in each internal point of the domain D_R. As the number ε is arbitrary, in each point (z,ξ) of the half-plane $z \geq 0$, $-\infty < \xi < \infty$, we have $w(z,\xi) = 0$, which was to be proved.

We introduce now the positive functions $\mu(z)$, $\overset{+}{v}(z)$, and $\overset{+}{\sigma}(z)$ as follows:

$$\tau(z) = \frac{\mathrm{d}}{\mathrm{d}z}\ln\mu(z) \quad (\mu(0)=1), \qquad \eta(z) = \frac{1}{\overset{+}{v}{}^2(z)}, \qquad \overset{+}{\sigma}(z) = \frac{\overset{+}{v}(z)}{\mu(z)}.$$

Set

$$f(\xi) = \frac{\mathrm{d}\varphi}{\mathrm{d}\xi} \qquad (\varphi(0) = 0);$$

and change the variables

$$x = \int_0^z \frac{\mathrm{d}z}{\overset{+}{v}(z)}, \qquad \sigma(x) = \overset{+}{\sigma}(z).$$

□

2.2. THE FORM OF SOLUTION OF THE DIRECT PROBLEM APPLIED FOR SOLUTION OF THE INVERSE PROBLEMS

Solution of the inverse problems from § 2.3 is based on the form of solution of the direct problem 2.1 which is given by Theorem 2.2.1.

Theorem 2.2.1. *Denote*

$$B(x) = -\frac{1}{2}\frac{\sigma'(x)}{\sigma(x)}; \quad q(x) = \sigma^{1/2}(x)\{\sigma^{-1/2}(x)\}'' = B' + B^2. \qquad (2.2.1)$$

Let the following conditions hold:

(A) *the derivatives* $\mathrm{d}^2\eta/\mathrm{d}z^2$ *and* $\mathrm{d}\tau/\mathrm{d}z$ *are continuous for* $0 \le z < \infty$;

(B) $\int_0^\infty x|q|\,\mathrm{d}x < \infty$, $\int_0^\infty B^2\,\mathrm{d}x < \infty$, $\int_0^\infty |B|\,\mathrm{d}x < \infty$ *(this holds, for example, if for a certain finite* $H > 0$ *we have* $\eta(z) \equiv \mathrm{const}$, $\mu(z) \equiv \mathrm{const}$ *for* $z \ge H$*)*;

(C) *the function* $\varphi(\xi)$ *and its derivatives up to the fourth order are continuous and absolutely integrable in the line* $-\infty < \xi < \infty$;

(D) $\tilde{\varphi}(k)|_{k=0} = 0$, *where* $\tilde{\varphi}(k)$ $(-\infty < k < \infty)$ *is the Fourier transform of the function* $\varphi(\xi)$

$$\varphi(\xi) = \int_{-\infty}^{+\infty} \tilde{\varphi}(k)e^{ik\xi}\,\mathrm{d}k. \qquad (2.2.2)$$

Then the solution $u(z,\xi)$ *of the direct problem 2.1 admits the representation*

$$u(z,\xi) = \int_{-\infty}^{\infty} U(z,k)e^{ik\xi}\,\mathrm{d}k, \qquad 0 \le z < \infty, \quad -\infty < \xi < \infty, \qquad (2.2.3)$$

$$U(z,k) = \sigma^{1/2}(z)W(x,k), \quad x = \int_0^z \frac{\mathrm{d}z}{\overset{+}{v}(z)}, \quad \sigma(x) = \overset{+}{\sigma}(z), \qquad (2.2.4)$$

$$W(x,k) = \begin{cases} \dfrac{\gamma ik\tilde{\varphi}(k)f(x,ik)}{f'(0,ik) - (\alpha + ik\beta)f(0,ik)}, & k \ge 0 \\[2ex] \dfrac{\gamma ik\tilde{\varphi}(k)\tilde{f}(x,ik)}{\tilde{f}'(0,ik) - (\alpha + ik\beta)\tilde{f}(0,ik)}, & k \le 0 \end{cases}$$

$$(0 \le x < \infty, \quad -\infty < k < \infty). \qquad (2.2.5)$$

Here

$$\alpha = B(0); \qquad \beta = \varkappa \cdot \sigma(0); \qquad \gamma = \sigma^{1/2}(0); \tag{2.2.6}$$

$\tilde{f}(x,k)$ *and* $f(x,k)$ *are defined in Subsection 1.2.2 solutions of integral equations* (1.2.6) *and* (1.2.11) *respectively (with the function* $q(x)$ *of the from* (2.2.1)).

The integral in formulas (2.2.3) and in the equality

$$\frac{\partial u}{\partial z} = \int_{-\infty}^{\infty} \frac{\mathrm{d}U}{\mathrm{d}z} e^{ik\xi}\, \mathrm{d}k$$

are the Fourier integrals (therefore, their inversion formulas hold).

Proof. The proof of this theorem is analogous to methods applied in § 1.2 and is based on applying Theorems 1.2.2 and 1.2.4.

The solution $u(z,\xi)$ of the direct problem 2.1 we seek in form (2.2.3) where the integral is taken along the real axis of the complex plane $k = k_1 + ik_2$. We suppose that the conditions (A) and (B) of Theorem 2.2.1 hold. Substituting the functions (2.2.2) and (2.2.3) into conditions of the direct problem 2.1 we change the variables (2.2.4). Then we pass to the new complex variable s by the formulas

$$s = ik, \quad W(x,k) = W(x,s). \tag{2.2.7}$$

As a result, we obtain the problem for the function $W(x,s)$

$$\frac{\mathrm{d}^2 W}{\mathrm{d}x^2} + \{s^2 - q(x)\}W = 0, \quad 0 \le x < \infty, \tag{2.2.8}$$

$$\frac{\mathrm{d}W}{\mathrm{d}x}(0,s) - (\alpha + s\beta)W(0,s) = \gamma ik\tilde{\varphi}(k). \tag{2.2.9}$$

The function $W(x,s)$ must be continuous and bounded in the imaginary axis $\operatorname{Re} s = 0$, $-\infty < \operatorname{Im} s < \infty$ and must decrease sufficiently fast for $|\operatorname{Im} s| \to \infty$, $\operatorname{Re} s = 0$ for each fixed $x \in [0,\infty)$[i)] (condition (2.2.10)).

The real function $q(x)$ and the number α, β, γ are expressed by formulas (2.2.1), (2.2.6).

Unlike system (1.2.8), (1.2.9), Chapter 1 for the hyperbolic equation, equation (2.2.8) and condition (2.2.9) are set not on the real axis, but on

[i)]The estimate for the function $W(x,s)$ sufficient for convergence of integral (2.2.3) and its derivatives depends on the conditions imposed on the function $\varphi(\xi)$ and is given below.

the imaginary axis $\operatorname{Re} s = 0$ of the complex plane. The solution $W(x, s)$ of problem (2.2.8)–(2.2.10) must be determined therefore for all purely imaginary values s and for $0 \le x < \infty$. Similarly the problem of Chapter 1 for the string equation, system (2.2.8), (2.2.9) does not correspond to self-adjoint Sturm—Liouville operator, since the coefficient $(\alpha + s\beta)$ in condition (2.2.9) is complex and depends on s.

By conditions (A), (B), the function $q(x)$ is continuous for $0 \le x < \infty$, and equation (2.2.8) has the solution $\tilde{f}(x, s)$ and $f(x, s)$ defined for $\operatorname{Im} s \le 0$ and $\operatorname{Im} s \ge 0$ respectively and satisfying integral equation (1.2.6) and (1.2.11) (their properties see in Subsection 1.2.2). Taking into account these functions we defined the solution $W(x, s)$ of system (2.2.8), (2.2.9) in the form

$$W(x,s) = \begin{cases} \dfrac{f(x,s)\gamma s\tilde{\varphi}(k)}{f'(0,s) - (\alpha + s\beta)f(0,s)}, & \operatorname{Im} s > 0, \\[2ex] \dfrac{\tilde{f}'(x,s)\gamma s\tilde{\varphi}(k)}{\tilde{f}'(0,s) - (\alpha + s\beta)\tilde{f}(0,s)}, & \operatorname{Im} s < 0, \end{cases} \tag{2.2.11}$$

$$\operatorname{Re} s = 0, \qquad x \ge 0.$$

In order that the function $W(x, s)$ of form (2.2.11) be continuous and bounded in the imaginary axis $\operatorname{Re} s = 0$ for $s \ne 0$ it is necessary and sufficient that the functions

$$A(s) = f'(0, s) - (\alpha - s\beta)f(0, s),$$

and

$$A_1(s) = f'(0, s) - (\alpha + s\beta)f(0, s)$$

have no zeros in the imaginary semiaxis $\operatorname{Re} s = 0$, $\operatorname{Im} s > 0$ of the plane $s = s_1 + is_2$. The lack of zeros of the functions $A(s)$ and $A_1(s)$ in the set $\operatorname{Re} s = 0$, $\operatorname{Im} s > 0$ which we define as D_0 is proved similarly Subsection 1.2.5.

Parallel with the function $A(s)$ ($s \in D_0$) we consider the set $\{L_{\theta_0}\}_{D_0}$ of not self-adjoint operators L_{θ_0}[ii)] dependent on the parameter $\theta_0 = \alpha - s_0\beta$, where $s_0 \in D_0$. We suppose, for definiteness, that $\beta > 0$ (i.e., $\varkappa > 0$). Then, for $s_0 \in D_0$, we have $\operatorname{Im} \theta_0 = -\operatorname{Im} s_0 \cdot \beta < 0$. By Theorem 1.2.2, the eigenvalues $\lambda = s^2$ of each operator from the set $\{L_{\theta_0}\}_{D_0}$ may lay only in the domain $\operatorname{Im} s^2 = 2s_1s_2 < 0$. In the half-plane $\operatorname{Im} s > 0$ the values s from the domain $s_2 > 0$, $s_1 < 0$ (the domain D_c in terms of Theorem 1.2.4) correspond

[ii)] Definition of the operator L_{θ_0} see in Subsection 1.2.4.

to these values $\lambda = s^2$. Since the sets D_0 and D_c are not intersecting, by Theorem 1.2.4, the function $A(s)$ has no zeros in the D_0. The analogous result holds and for $\beta < 0$.

Analogously, considering the set of operators $\{L_{\theta_0}\}_{D_0}$ dependent on the parameter $\theta_0 = \alpha + s_0\beta$, where $s_0 \in D_0$, we can show that the function $A_1(s)$ has no zeros in the set D_0 for $\beta > 0$ and $\beta < 0$.

If $\beta = 0$, then $A(s) = A_1(s)$ and the operator L_{θ_0} with the parameter $\theta_0 = \alpha$ is self-adjoint. It is positive definite, which follows from Alekseev (1962); therefore, it has no negative eigenvalues[iii)]. This implies that for $\beta = 0$ (i. e., for $\varkappa = 0$) there are no zeros of the functions $A(s)$ and $A_1(s)$ in the set D_0.

Further, by the properties of the numbers z_1 and y_1, from Theorem 1.2.6 it follows that, under the conditions (A), (B), the limits

$$\lim_{s\to 0} A_1(s)/s = -\beta y_1 - iz_1; \qquad \lim_{s\to 0} \tilde{A}(s)/s = -\beta y_1 + iz_1,$$

where

$$\tilde{A}(s) = \tilde{f}'(0,s) - (\alpha + s\beta)\tilde{f}(0,s) = A(-s) \qquad (\operatorname{Im} s \le 0)$$

are finite, distinct and not equal to zero. So, the function $W(x, s)$ has a finite gap with respect to s for each $x \ge 0$ if $\tilde{\varphi}(k)|_{k=0} \ne 0$.

Let $\tilde{\varphi}(k)|_{k=0} = 0$, i. e., the condition (D) of Theorem 2.2.1 holds (it holds, for example, if $\varphi = \mathrm{d}\varphi_J/\mathrm{d}\xi$, where the function $\varphi_J(\xi)$ is absolutely integrable in the axis $-\infty < \xi < \infty$). Then, the function $W(x, s)$ is continuous in the whole line $\operatorname{Re} s = 0$, for $x \ge 0$.

Returning by formulas (2.2.7) to the complex plane k and using the relations

$$W(x,-k) = \bar{W}(x,k), \qquad |W(x,k)| < Me^{-|k|x} \cdot |\tilde{\varphi}(k)|,$$
$$-\infty < k < \infty, \qquad 0 \le x < \infty$$

and the known relation between the smoothness of $\varphi(\xi)$ and decreasing $\tilde{\varphi}(k)$ (see Subsection 1.2.8) for $|k| \to \infty$, we obtain Theorem 2.2.1. Simultaneously, we have shown the existence of solution of the direct problem 2.1. This solution, as it have been shown above, is unique. □

[iii)] The proof of the positive definite property of the operator L_{θ_0} in this case is obtained from the establishment of the analogous statement in Subsection 1.2.6 if, in this statement, we set $\beta = 0$.

2.3. THE SETTING AND SOLUTION OF THE INVERSE PROBLEMS

Inverse problem 2.1. We suppose that for $-\infty < \xi < \infty$ the values $u(0,\xi)$ and $(\partial u)/(\partial z)|_{z=0}$ are given, where $u(z,\xi)$ is the solution of the direct problem 2.1 with a certain unknown functions $\eta(z)$, $\tau(z)$ $(0 \le z < \infty)$, $f(\xi)$ $(-\infty < \xi < \infty)$ and an unknown number $\varkappa \in (-\infty, \infty)$.

We need to find: *in the general case* — the function $\sigma(x)$ for $0 \le x < \infty$; if $\tau(z) \equiv 0$, then the function $\eta(z)$ for $0 \le z < \infty$; if the function $\eta(z)$ is known for $0 \le z < \infty$, then — the function $\tau(z)$ for $0 \le z < \infty$.

Parallel with differential system (2.2.8), (2.2.9) set in the imaginary axis of the plane s, we consider the self-adjoint Sturm—Liouville system

$$\frac{\mathrm{d}^2 y}{\mathrm{d}x^2} + \{s^2 - q(x)\}y = 0, \qquad 0 \le x < \infty, \tag{2.3.1}$$

$$y(0,s) = 1, \qquad y'(0,s) = \alpha, \tag{2.3.2}$$

being set of real s; the function $q(x)$ and the number α here are taken from system (2.2.8), (2.2.9).

Similarly the inverse problem 1.1 for the string, we reduce the inverse problem 2.1 for elliptic equation (2.1.1) to the inverse Sturm—Liouville problem. This means that we find from the data of the inverse problem 2.1 either the main transition function

$$\Phi(t) = \int_0^\infty \frac{1 - \cos\sqrt{\lambda}t}{\lambda}\,\mathrm{d}\rho(\lambda)$$

of system (2.3.1), (2.3.2) for $t \ge 0$, or its spectral function $\rho(\lambda)$ from the formula

$$\mathrm{d}\rho(\lambda) = \frac{2s^2\,\mathrm{d}s}{\pi|f'(0,s) - \alpha f(0,s)|^2}, \qquad \lambda = s^2, \quad -\infty < s < \infty$$

for $\lambda \ge 0$ $(\rho(0) = 0)$. However, this inverse problem for elliptic equation (2.1.1) differs essentially from the inverse problem 1.1 for string by the method of deriving these functions. The reason is that although the integral (2.2.3) (and in formula (1.2.2)) we calculate along the real axis $\operatorname{Im} k = 0$ of the plane $k = k_1 + ik_2$, the values of the functions $f(x,k)$ and $\tilde{f}(x,k)$ entering in $W(x,k)$ are taken for purely imaginary values $k = ik_1$, which is seen from formula (2.2.5).

Therefore, inversion of the Fourier integrals for $u(0,\xi)$ and $\partial u/\partial z|_{z=0}$ and the formulas (we use Theorem 2.2.1).

$$\frac{ikU(0,k)}{\mathrm{d}U/\,\mathrm{d}z|_{z=0}} = \frac{\sigma(0)sf(0,s)}{f'(0,s)-\alpha f(0,s)} = F(s), \qquad \operatorname{Im} s \geq 0 \quad (\operatorname{Re} k \geq 0), \tag{2.3.3}$$

$$\frac{ikU(0,k)}{\frac{\mathrm{d}U}{\mathrm{d}z}|_{z=0}} = \frac{\sigma(0)\tilde{f}(0,s)s}{\tilde{f}'(0,s)-\alpha\tilde{s}(0,s)} = \tilde{F}(s), \qquad \operatorname{Im} s \leq 0 \quad (\operatorname{Re} k \leq 0) \tag{2.3.4}$$

$$s = ik, \qquad \operatorname{Re} s = 0$$

give the values of functions $F(s)$ and $\tilde{F}(s)$ by means of which we construct the function $\rho(\lambda)$ (see § 1.3)

$$\mathrm{d}\rho(\lambda) = -\frac{1}{\sigma(0)}\frac{i}{\pi}\{\tilde{F}(s)-F(s)\}\,\mathrm{d}s, \qquad -\infty < s < \infty \tag{2.3.5}$$

not in the real axis $-\infty < s < \infty$, but in the imaginary semiaxes $s_2 \geq 0$ and $s_2 \leq 0$ respectively.

The function $F(s)$ defined by formula (2.3.3) is analytic in the half-plane $\operatorname{Im} s > 0$ and continuous up to the real axis $\operatorname{Im} s = 0$ (this follows from the properties of the function $f(x,s)$ and the statement of Subsection 1.2.6). Therefore, by the uniqueness theorem for analytic functions, $F(s)$ is determined uniquely everywhere in the half-plane $\operatorname{Im} s > 0$; in particular, in the certain line $\operatorname{Im} s = b_0 > 0$. We extend analytically the function $F(s)$ defined by formula (2.3.3) from the extremely small segment $[i(b_0-\Delta), ib_0]$ (the numbers b_0 and $\Delta < b_0$ are positive, arbitrary, and finite) of the imaginary semiaxis $s = is_2$, $s_2 \geq 0$ to the whole line $\operatorname{Im} s = b_0$ of the plane $s = s_1 + is_2$ using the Taylor expansion

$$F(s) = \sum_{n=0}^{\infty} \frac{F^{(n)}(s_k)}{n!}(s-s_k)^n, \qquad k = 0,1,2,\dots$$

along the chain of closed disks

$$K_0 : |s-s_0| \leq r_0; \quad K_1 : |s-s_1| \leq r_0; \quad \dots; \quad K_k : |s-s_k| \leq r_0.$$

The radius is $r_0 = b_0 - \varepsilon$, $0 < \varepsilon < b_0$ (ε is a small positive number); the centers

$$s_0 = ib_0, \quad s_1 = \pm(b_0-\varepsilon) + ib_0, \dots, s_k = \pm k(b_0-\varepsilon) + ib_0$$

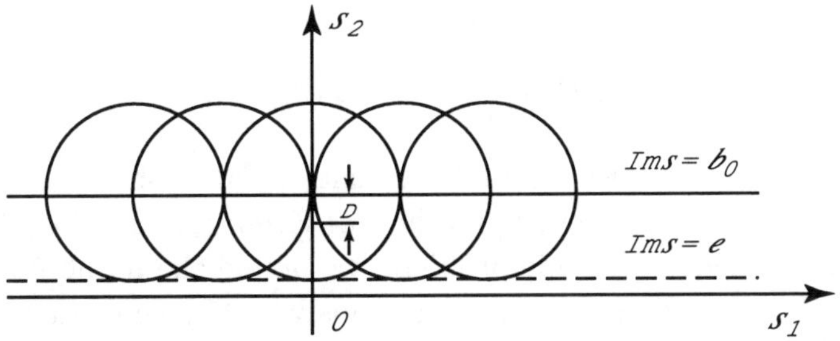

Figure 2.1. To the solution of the inverse problem for the elliptic equation in the half-plane

of these disks we move subsequently to the right and to the left from the center $s_0 = ib_0$ of the first disk K_0 along the line $\operatorname{Im} s = b_0$ (Figure 2.1). The number b_0 (and, therefore, the radius r_0) we may take extremely large (but finite).

The function $F(s)$ in the real axis we do not define following this scheme; therefore, by formula (2.3.5), where $\tilde{F}(s) = -F(-s)$, we cannot find (as in the inverse problem of the string in Chapter 1) the spectral function $\rho(\lambda)$ for $\lambda = s^2 \geq 0$. Instead of the function $\rho(\lambda)$ we define the principal transition function $\Phi(t)$ of system (2.3.1), (2.3.2) in any finite interval $0 \leq t \leq T$.

To this end we consider the function

$$H(s,t) = \frac{\sin st}{s} \frac{i}{\pi} F(s) = \sigma(0) \frac{i}{\pi} \frac{\sin st f(0,s)}{f'(0,s) - \alpha f(0,s)}.$$

Evidently, in the half-plane $\operatorname{Im} s > 0$, the function $H(s,t)$ is analytic with respect to s for each fixed $t \in [0,T]$; it is continuous up to the real axis $\operatorname{Im} s = 0$ and is known in the whole line $\operatorname{Im} s = b_0$.

Therefore, from the asymptotic properties of the function $f(x,s)$ (see Subsection 1.2.2) and the Cauchy theorem it follows that

$$\int_{-\infty}^{\infty} H(s,t)\, \mathrm{d}s = \int_{-\infty+ib_0}^{\infty+ib_0} H(s,t)\, \mathrm{d}s$$

in each finite interval $0 \leq t \leq T$ (we integrate along the lines $\operatorname{Im} s = b_0$ and $\operatorname{Im} s = 0$).

Note that

$$\int_{-\infty}^{\infty} H(s,t)\, \mathrm{d}s = \sigma(0), \qquad \Phi'(t) = \sigma(0) \int_0^{\infty} \frac{\sin st}{s}\, \mathrm{d}\rho(\lambda).$$

Taking into account the properties

$$\lim_{t\to 0}\Phi'(t)=1, \qquad \Phi(0)=0,$$

we find the value $\sigma(0)$ and the main transition function $\Phi(t)$ in each finite interval $0\le t\le T$. Applying the method of Krein (1953, 1954) for solution of the inverse Sturm—Liouville problem, we define the number α, the function $q(x)$ and then (for example, analogously § 1.3) the function $\sigma(x)$ for $0\le x<\infty$. The further solution of the inverse problem, when $\tau(z)\equiv 0$, or when $\overset{+}{v}(z)$ is set for $0\le z<\infty$, is analogous to those from § 1.3.

From this way of solution of the inverse problem 2.1 the theorem follows.

Theorem 2.3.1. *We suppose that the condition (A), (B), (C), (D) of Theorem 2.2.1 hold. Then the inverse problem 2.1 has a unique solution for any real* $\varkappa$.

Remark 2.3.1. Let for a certain $\varepsilon>0$,

$$\int_0^\infty e^{\varepsilon x}|q(x)|\,dx<\infty. \tag{B_ε}$$

Then the simpler method for solution of the inverse problem 2.1 may be suggested. It is based on the following lemma.

Lemma 2.3.1. *If the condition (B_ε) holds, then the function $F(s)$ is analytic in the half-plane $\operatorname{Im}s>-\delta$, where δ is a certain positive and fixed number ($\delta\le 1/2\varepsilon$). So, all points of the real axis are internal points of the domain where $F(s)$ is analytic (regular points of the function $F(s)$).*

Proof. If the condition (B_ε) holds, then, by Theorem 2.6.1 (see, Naymark, 1954), the function $f(x,s)$ for each value $x\in[0,\infty)$ is analytic with respect to s in the half-plane $\operatorname{Im}s>-1/2\varepsilon$. Therefore, the function

$$A_0(s)=f'(0,s)-\alpha f(0,s)$$

is analytic also when $\operatorname{Im}s>-1/2\varepsilon$. As, for $\beta=0$, we have $A(s)=A_1(s)=A_0(s)$, then zeros of the function $A_0(s)$ may be displaced only in the strip

$$-\varepsilon/2<\operatorname{Im}s<0, \qquad -\infty<\operatorname{Re}s<\infty. \tag{2.3.6}$$

By Theorem 3.3.2 (see, Naymark, 1954), the zeros of the function $A_0(s)$ form a bounded set in the strip (2.3.6). Since the limit points of the set of zeros of the function $A_0(s)$ may lay only in the line $\operatorname{Im} s = -\varepsilon/2$, then the function $A_0(s)$ has a finite number of zeros in each half-plane $\operatorname{Im} s \geq -\varepsilon'$, where $0 < \varepsilon' < \varepsilon/2$ (for example, $\varepsilon' = \varepsilon/4$). Moreover, all these zeros (except for $s = 0$) lie in a certain rectangle

$$-\varepsilon' \leq \operatorname{Im} s < 0, \quad -a < \operatorname{Re} s < a, \quad 0 < a < \infty \tag{2.3.7}$$

from the strip (2.3.6). Therefore, there exists a positive number δ, so that the function $A_0(s)$ will have no zeros in the strip

$$-\delta \leq \operatorname{Im} s < 0, \qquad -\infty < \operatorname{Re} s < \infty$$

(except for $s = 0$), whence Lemma 2.3.1 follows. □

Hence it follows that if (B_ε) holds we may extend the function $F(s)$ analytically in the real axis $-\infty < s < \infty$ using the Taylor expansion for the function $F(s)$. Then, by formula (2.3.6), similarly as in Chapter 1, we find either the spectral function $\rho(\lambda)$ of system (2.3.1), (2.3.2) for all $s \in (-\infty, \infty)$, or the main transition function $\Phi(t)$ for $t \in [0, \infty)$. After, we solve the inverse Sturm—Liouville problem similar to the case of the hyperbolic equation of Chapter 1, defining the function $\sigma(x)$ for $0 \leq x < \infty$.

We formulate now another inverse problem which (unlike the inverse problem 2.1) has the meaning only for $\varkappa \neq 0$ and has no analog in the case $\varkappa = 0$.

We suppose that the value $\mu(0)$ is arbitrary (and is set in the direct problem 2.1). We introduce the number $\delta > 0$ and the function $\varphi^*(\xi)$ by the formulas

$$\varkappa = \delta/\mu(0), \qquad \varphi(\xi) = \varkappa\varphi^*(\xi), \qquad \varkappa \neq 0.$$

Then Theorem 2.2.1 holds true if in (2.2.5), instead of $\tilde{\varphi}(k)$, we substitute the function $\tilde{\varphi}^*(k)$ — the Fourier transform of the function $\varphi^*(\xi)$ and replace formulas (2.2.6) for the numbers β and γ by expression

$$\beta = \delta\sigma(0), \qquad \gamma = \beta\sigma^{-1/2}(0).$$

Inverse problem 2.2. Let, for $-\infty < \xi < \infty$, the function $\varphi^*(\xi)$ and the values $u(0, \xi)$ of the solution $u(z, \xi)$ of the direct problem 2.1 be set. The functions $\eta(z)$, $\mu(z)$ $(0 \leq z < \infty)$ and the number $\varkappa \neq 0$ are unknown, and the number $\delta \neq 0$ is known.

We need to define: in the general case — the function $\sigma(x)$ for $0 \leq x < \infty$; if $\mu(z) \equiv \text{const} = \mu$ and the number μ is known, then — the function $\eta(z)$ for $0 \leq z < \infty$; if $\eta(z)$ is known, then — the function $\mu(z)$ for $0 \leq z < \infty$.

This inverse problem we reduce to the problem of determination of differential system (2.3.1), (2.3.2) (analogously the inverse problem 2.1) by the main transition or spectral function[iv] and prove the following theorem.

Theorem 2.3.2. *Suppose that all the conditions of Theorem 2.2.1 hold. Then the inverse problem 2.2 has a unique solution.*

[iv]However, the function $\Phi(t)$ and the value $\sigma(0)$ are defined somewhat differently. These distinctions are analogous to those operations from § 1.4.

Chapter 3.

Inverse problems of scattering plane waves from inhomogeneous transition layers (half-space)

This chapter is devoted to the problems of determining the characteristics of inhomogeneous elastic layer (half-space) given the forms of incident and reflected waves.

Note that namely direct and inverse problems of scattering plane waves from inhomogeneous transition layers (half-space) were the reason of setting the direct and inverse problems considered in Chapters 1 and 2. The result of these chapters are immediately transformed on the problems considered in Chapter 3 (§ 3.1, 3.2). The problems of this chapter are the physical applications relatively the problems of Chapters 1, 2.

However, in the problem with plane waves we formulate and investigate in § 3.3, 3.4 a new inverse problem different from the problems considered in Chapters 1, 2. The possibility of considering the new settings in connected with a new additional parameter (the incidence angle θ_0 of the plane wave). The solution of the direct problem is dependent on this parameter.

As was noted in Introduction, in investigations of inverse problems of theory of wave propagation the following problem is of great interest. From homogeneous half-space $z < 0$, a longitudinal (transverse) wave of the form $\varphi_0(\xi)$ is normally falling, generating the reflected plane wave of the form $\varphi_1(\xi)$. The wave is falling on the inhomogeneous elastic half-space

$z \geq 0$ (or, in particular, on the transition layer). The properties of this half-space — the density ρ and the speed v of propagating longitudinal (transverse) waves are arbitrary functions of the normal coordinate z (depth). We need to find the elastic properties of the half-space (layer) given the functions $\varphi_0(\xi)$ and $\varphi_1(\xi)$.

The various methods and numerical algorithms for solving this inverse problem are suggested by Nikolskii (1964), Borodaeva (1969), Mikhailova and Parijskii (1969).

It is essential that in this problem we may recover this functions $\rho(z)$ and $v(z)$ not exactly, but a certain function $\sigma_0(x)$, which is obtained from the function $\sigma_0(z) = \rho(z)v(z)$ by the change of variables

$$x = \int_0^z \frac{dz}{v(z)}, \qquad \sigma_0(x) = \sigma_0(z).$$

Thus, using the information given in this inverse problem, we can recover only acoustic shiftness $\sigma_0(x)$ as the function dependent on time x of wave run along the segment $[0, z]$. The time of running $x(z)$ and the properties $\rho(z)$, $v(z)$ as functions of depth are remained unknown.

In this connection, the following questions are of the interest.

First, more complete determination of the elastic properties of the inhomogeneous layer; for example, the functions $v(z)$ and $\rho(z)$.

Second, research of the inverse problems for the general case, when the plane wave is falling at the arbitrary angle.

This chapter is devoted to solution of this two questions (these questions are closely connected) in the case when the incident wave is the transverse wave of the SH type.

Note that all the below discussions, theorems and formulas are carried over on the acoustic case (see Remark 3.5 below).

In § 3.1 we show that the problem of scattering the plane waves SH falling obliquely from the inhomogeneous elastic half-space (layer) is reduced to the certain two-dimensional problem in the half-plane with directional derivative in the boundary condition. However, depending on the characteristics inhomogeneous medium and the angle of incidence of the plane wave, the differential equation of this problem may have hyperbolic, elliptic or mixed type.

In this chapter we consider the inverse problems of hyperbolic and elliptic cases. The hyperbolic case generalizes the problem of normal incidence. The elliptic case is differed fundamentally from the hyperbolic case and corresponds to the phenomena of complete internal reflection.

In both cases we, first of all, shall formulate the problem on establishment of properties of the inhomogeneous layer by the forms of incident and reflected waves given for a single value of the angle θ_0. The uniqueness theorems and the way of solution of such inverse problem are obtained by simple transferring the results of Chapters 1 and 2. In the general case, we can recover uniquely a certain intermediate function $\sigma(x, \theta_0)$ (which coincide with the function $\sigma_0(x)$ when $\theta_0 = 0$). The functions $v(z)$ and $\rho(z)$ are not determined.

In this connection, in § 3.3, we suggest the way of determination of the density and the speed in the inhomogeneous layer as the functions of the depth given the sets of plane waves reflected from the layer at various angles. The functions $v(z)$ and $\rho(z)$ are recovered both in hyperbolic and in elliptic cases if we set the forms $\{\varphi_0(\xi, \theta_0), \varphi_1(\xi, \theta_0)\}$ of incident and reflected waves for a certain set of the values of θ_0. Such set may be, for example, an arbitrary converging sequence. In § 3.3 the uniqueness theorem is established and the way of solution of this problem is described[i)]. The method of solution is constructive and is expressed in the calculated algorithm (§ 3.4).

In other words, in this chapter we suggest also the numerical algorithm of determination of the density and the velocity in the layer, as the functions of the depth. The results of numerical calculations are shown for the hyperbolic case (as was noted, the results of Nikolskii (1964), Borodaeva (1969), Mikhailova and Parijskii (1969) are devoted to determination of a certain function $\sigma_0(x)$ but not of the functions $v(z)$, $\rho(z)$).

The method suggested here divides the solution of recovery of the functions $\rho(z)$ and $v(z)$ onto two stages. First, we solve, for each fixed θ_0 from the sequence Θ_0, the problem of determination a certain intermediate function $\sigma(x, \theta_0)$ (coincided with the function $\sigma_0(x)$ for $\theta_0 = 0$) by the two functions $\varphi_0(\xi, \theta_0)$, $\varphi_1(\xi, \theta_0)$. As a result, we find a sequence of functions $\sigma(x, \theta_0)$, where $\theta_0 \in \Theta_0$. At last, we consider the final solution $\{v(z), \rho(z)\}$ of the inverse problem using an explicit formula which contains the value of the derivative $\partial\sigma(x, \theta_0)/\partial\theta_0$ in the limit point. For this purpose, we need to know the values of $\sigma(x, \theta_0)$ in a sequence of points converging to $\bar{\theta}_0$.

It should be noted that in order to calculate $\partial\sigma/\partial\theta_0$ numerically it suffices to know the values $\sigma(x, \theta_0)$ in two (or three) points θ_0, then, in order to

---

[i)] The possibility of determination both $v(z)$ and $\rho(z)$ when solving the inverse dynamic problems of seismics (which is inaccessible for kinematic methods) was noticed by Alekseev (1962). In this paper the both functions $v(z)$ and $\rho(z)$ are recovered in inhomogeneous space excited by the point source. The vibration field is setted on the whole boundary of the half-space (i. e., in the plane).

determine $v(z)$ and $\rho(z)$ numerically we need to know the inverse problem data (the functions $\varphi_0(\xi, \theta_0)$, $\varphi_1(\xi, \theta_0)$) for two (or three) close angles θ_0, $\theta_0 + \Delta\theta_0$ where $\Delta\theta_0$ is small.

In this connection, the author had suggested that for uniqueness of solution of the inverse problem on recovery the two characteristics $v(z)$ and $\rho(z)$ of the media it suffices to know the function $\sigma(x, \theta_0)$ (i. e., to set the inverse problem data $\{\varphi_0(\xi, \theta_0), \varphi_1(\xi, \theta_0)\}$) also for two different angles θ_0. Such inverse problem was formulated by the author and was suggested for research to V. A. Gorbunov. It occurs that really, the functions $v(z)$ and $\rho(z)$ are determined uniquely given the two functions $\sigma(x, \theta_0)$, i. e., the two collections $\{\varphi_0(\xi, \theta_0), \varphi_1(\xi, \theta_0)\}$ set for two different angles θ_0^1, θ_0^2. This result was obtained by Dobrinskii and Gorbunov (1977) for piecewise analytic functions $\rho(z)$ and $v(z)$.

Further, Blagoveshchenskii and Voevodskii (1981) had researched also such settings of the inverse problems of scattering the plane waves on inhomogeneous half-space in the hyperbolic case. In this case, the reflected waves are set for finite number of incidence angles.

Note that the results of § 3.3, in particular, formulas (3.3.3) and (3.5.1) hold not only for the problems with the transition layer $z \geq 0$ (half-space), but for the problem with the layer $0 \leq z \leq H$ with the fixed boundary $z = H$ or with the boundary free from stresses (it corresponds to the day surface in seismology). The reason is that the formulas obtained below do not depend on the shape of the domain where the differential equation is considered and on the boundary conditions. These formulas express only the connection between the coefficients in equation and the function $\sigma(x, \theta_0)$. This transfer of results of § 3.3 on the problem with free or fixed boundary is realized in § 6.3 (see the inverse problem 6.3).

Note that in order to know the functions $\varphi_0(\xi, \theta_0)$ and $\varphi_1(\xi, \theta_0)$ for a certain value θ_0 it suffices to measure the incident and reflected waves in a certain fixed point of the domain $z \leq 0$ as the function of time t.

Remark that for various values θ_0 the form $\varphi_0(\xi)$ of the incident impulse may be various: $\varphi_0 = \varphi_0(\xi, \theta_0)$. In this case, it is not necessary that the function $\varphi_0(\xi, \theta_0)$ be dependent on θ_0 smoothly (for example, continuously). It may be discontinuous with respect to θ_0.

All this holds true both for the hyperbolic case ($v(z) < v_0/\sin\theta_0$) and for the elliptic case ($v(z) > v_0 \sin\theta_0$). We are speaking about the scattering the plane waves from inhomogeneous transition layers (half-space) and from the layers with a free (fixed) boundary.

Everywhere in the inverse problems of this chapter the conditions imposed on the functions $\rho(z)$ and $v(z)$ being sufficient for uniqueness of solution of these problems are similar to those imposed on coefficients $\mu(z)$, $\eta(z)$, $\tau(z)$ from Chapters 1 and 2. These conditions are the smoothness conditions. In this case, the characteristics of the medium — the functions $v(z)$ and $\rho(z)$ — may be not monotone.

3.1. THE DIRECT PROBLEM

Given the two half-spaces $z \leq 0$ and $z \geq 0$ completed by elastic isotropic medium with the Lame coefficients λ, μ and the density ρ. The half-space $z < 0$ is homogeneous ($\lambda = \lambda_0$, $\mu = \mu_0$, $\rho = \rho_0$ are constant) and the space $z > 0$ is inhomogeneous. In this half-space the functions $\lambda = \lambda(z)$, $\mu = \mu(z)$, $\rho = \rho(z)$ are arbitrary functions dependent on z whose smoothness will be discussed below.

From the homogeneous medium $z < 0$, on the boundary $z = 0$, the plane wave of SH type is falling whose shift vector $\vec{u}^{\,0}(x, y, z, t)$ has the following projections:

$$u_x^0 \equiv 0, \quad u_z^0 \equiv 0, \quad u_y^0 = \varphi_0(t - (x \sin\theta_0 + z\cos\theta_0)/v_0).$$

Here θ_0 is the angle between the z-axis and the vector normal to the wave front set. The y axis is taken normal to the wave incidence plane. The origin of (x, z)-coordinate system is taken in a certain boundary point $z = 0$. The x-axis is disposed along the boundary in the incidence plane with the positive direction turned to propagation of the primary wave (Figure 3.1).

The initial time moment is the moment when the leading front set of the following wave achieves the point $x = 0$, $z = 0$. The positive time direction t corresponds to transition of the point of intersection of the wave front set with the boundary $z = 0$ in the positive direction of the x-axis.

The complete shift vector $\vec{u}$ in the problem has the only nontrivial component $u_y(x, z, t)$ which we denote as $u_0(x, z, t)$ in the half-plane $z \leq 0$, and $u(x, z, t)$ in the half-plane $z \geq 0$. From the system of equations of elastic theory (see, Leibenzon, 1942), we can obtain the wave equation for the function $u_0(x, z, t)$ and the equation of wave type for the function $u(x, z, t)$.

So, we consider the following problem in the space x, z, t.

Let, for $z < 0$, we have the equation

$$\frac{\partial^2 u_0}{\partial x^2} + \frac{\partial^2 u_0}{\partial z^2} = \frac{1}{v_0^2}\frac{\partial^2 u_0}{\partial t^2}; \qquad v_0 = \text{const}, \tag{3.1.1}$$

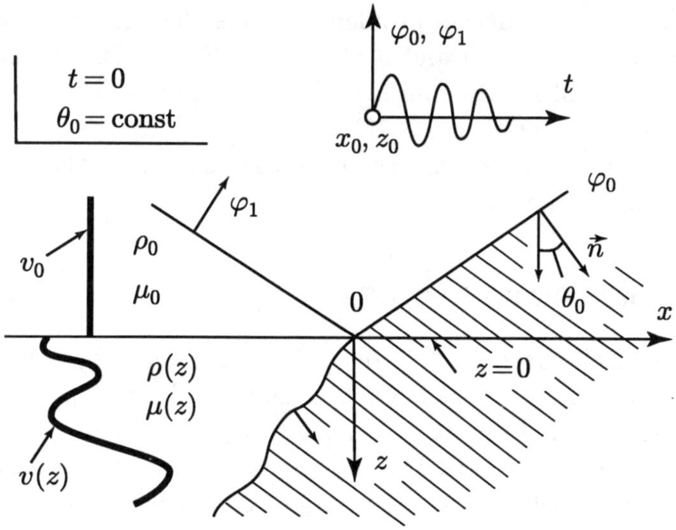

Figure 3.1. To the problem of scattering the plane waves from inhomogeneous transition layers (half-space) in the coordinates x, z, t

and, for $z > 0$, we have the equation

$$\frac{\partial^2 u}{\partial x^2} + \frac{\partial^2 u}{\partial z^2} + \frac{\mathrm{d}}{\mathrm{d}z}\{\ln \mu(z)\}\frac{\partial u}{\partial z} = \frac{1}{v^2(z)}\frac{\partial^2 u}{\partial t^2}. \tag{3.1.2}$$

We see that this last equation contains in the inhomogeneous medium, for $\mu(z) \not\equiv$ const the term with the first derivative of $u(x, z, t)$ with respect to z.

We set the conditions (of continuity of shifts and stresses) on the boundary $z = 0$

$$u(x, 0, t) = u_0(x, 0, t), \qquad \mu(0)\frac{\partial u}{\partial z}\Big|_{z=0} = \mu_0\frac{\partial u_0}{\partial z}\Big|_{z=0}. \tag{3.1.3}$$

We suppose in conditions (3.1.1)–(3.1.3) that $-\infty < x, t < \infty$. The constants v_0, μ_0 and the functions $v(z)$, $\mu(z)$ are positive. Each of the conditions $v(0) = v_0$, $v(0) \neq v_0$; $\mu(0) = \mu_0$; $\mu(0) \neq \mu_0$ may be true.

We suppose that the plane wave of the form $\varphi_0(\xi)$ $(-\infty < \xi < \infty)$ is falling from the half-plane $z < 0$ at the angle θ_0 $(0 \leq \theta_0 \leq \pi/2)$ on the boundary $z = 0$. It generates the plane wave $\varphi_1(x, z, t)$ reflected from the half-plane $z \geq 0$. This denotes that the function $u_0(x, z, t)$ is represented in the form

$$u_0(x, z, t) = \varphi_0\Big(t - \frac{x\sin\theta_0 + z\cos\theta_0}{v_0}\Big) + \varphi_1\Big(t - \frac{x\sin\theta_0 - z\cos\theta_0}{v_0}\Big), \tag{3.1.4}$$

where the function $\varphi_1(\xi)$ $(-\infty < \xi < \infty)$ — the form of reflected wave — is unknown. We suppose that

$$\varphi_0(\xi)|_{\xi<0} = 0, \tag{3.1.5}$$

i. e., the incident wave has the leading front.

We consider the case, when

$$v(z) < v_0/\sin\theta_0, \qquad 0 \le z < \infty, \tag{H}$$

and introduce the following condition for $t = 0$ (Figure 3.1):

$$u(x, z, 0) \equiv 0 \qquad \text{if} \quad x > 0, \quad z \ge 0. \tag{3.1.6}$$

In the case (H), we consider the following direct problem: the constants μ_0, v_0, and θ_0 $(0 \le \theta_0 < \pi/2)$ are given; the functions $v(z)$ and $\mu(z)$ for $0 \le z < \infty$ are given, and the function $\varphi_0(\xi)$ is given for $-\infty < \xi < \infty$ satisfying property (3.1.5). We need to find the functions $u(x, z, t)$ $(z \ge 0)$, $u_0(x, z, t)$ $(z \le 0)$, $\varphi_1(\xi)$, $(-\infty < \xi < \infty)$, satisfying conditions (3.1.1)–(3.1.6) for given θ_0.

We suppose below that the function φ_0 may depend on θ_0: $\varphi_0 = \varphi_0(\xi, \theta_0)$.

From formula (3.1.4) it follows that the summary field $u_0(x, z, t)$ on the boundary $z = 0$ is the function dependent on the only variable $\xi = t - x\sin\theta_0/v_0$. Therefore, it is natural to seek the field $u(x, z, t)$ in the half-plane $z \ge 0$ as a function of two variables z and ξ: $u(x, z, t) = u(z, \xi)$.

Three-dimensional problem (3.1.1)–(3.1.6) on determining the three functions u, u_0, φ_1 can be reduced to the two-dimensional problem dependent on the variables z, and $\xi = t - x\sin\theta_0/v_0$: find the function

$$u(z, \xi) = u(z, \xi, \theta_0) = u(x, z, t, \theta_0)$$

when the equation

$$\frac{\partial}{\partial z}\left\{\mu(z)\,\frac{\partial u}{\partial z}\right\} = \eta(z)\,\frac{\partial^2 u}{\partial \xi^2} \tag{3.1.7}$$

is given in the half-plane $z > 0$, $-\infty < \xi < \infty$. Here the function

$$\eta(z) = \frac{\mu(z)}{\bar{v}^2(z, \theta_0)}, \qquad \frac{1}{\bar{v}^2(z, \theta_0)} = \frac{1}{v^2(z)} - \frac{\sin^2\theta_0}{v_0^2} \tag{3.1.7'}$$

is positive by the condition (H) for $0 \le z < \infty$.

The boundary condition

$$\left(\frac{\partial u}{\partial z}-\varkappa\frac{\partial u}{\partial \xi}\right)\Big|_{z=0}=\frac{\partial \varphi}{\partial \xi}, \tag{3.1.8}$$

is given for $z=0$, $-\infty<\xi<\infty$. Here

$$\varkappa=\frac{\delta}{\mu(0)}, \qquad \delta=\cos\theta_0\,\frac{\mu_0}{v_0}, \qquad \varphi(\xi)=-2\varkappa\varphi_0(\xi,\theta_0). \tag{3.1.8$'$}$$

Besides, the initial condition is given

$$u(z,\xi,\theta_0)|_{\xi<0}\equiv 0, \qquad z\geq 0. \tag{3.1.9}$$

Equality (3.1.8) is the condition of the directional derivative type. It is obtained by elimination of the unknown function $\varphi_1(\xi,\theta_0)$ from the conditions

$$u(0,\xi,\theta_0)=\varphi_0(\xi,\theta_0)+\varphi_1(\xi,\theta_0), \tag{3.1.10}$$

$$\frac{\partial u}{\partial z}\Big|_{z=0}=\frac{\cos\theta_0}{v_0}\frac{\mu_0}{\mu(0)}\Big\{-\frac{\mathrm{d}\varphi_0}{\mathrm{d}\xi}+\frac{\mathrm{d}\varphi_1}{\mathrm{d}\xi}\Big\}, \tag{3.1.11}$$

equivalent to the conditions (3.1.3).

Initial equation (3.1.1) was hyperbolic. However, equation (3.1.7) may be hyperbolic, elliptic or mixed depending on the sign of function (3.1.7$'$). Only in the particular case of normal fall of the plane wave, when $\theta_0=0$, equation (3.1.7) is always hyperbolic.

The case (H) is naturally called hyperbolic, since, under the condition (H), equation (3.1.7) is hyperbolic in the half-plane $z\geq 0$, $-\infty<\xi<\infty$.

It is essential that in the hyperbolic case, the value $v(z)$ may be both less and greater than v_0. This case holds for any finite values v_0, $v(z)$ $(0\leq z<\infty)$ for sufficiently small incidence angles θ_0 of the wave φ_0.

Remark 3.1.1. Condition (3.1.6) necessarily holds in the case of normal incidence of the wave φ_0, when $\xi=t$. It holds also in the case (H) for $v(z)\equiv\text{const}$, when the solution $u(x,z,t)$ is searched in the form of a plane wave. In the general case this condition must be formulated and setted explicitly. This condition follows from the assumption on existence of leading front of the wave $u(x,z,0)$ in the half-space $z\geq 0$.

Problem (3.1.7)–(3.1.9), evidently, is a certain case of the boundary-value problem 1.1 for the semibounded string. If the smoothness conditions imposed on the function $u(z,\xi) = u(z,\xi,\theta_0)$ in § 1.1 hold, direct problem (3.1.7)–(3.1.9) (and, therefore, initial problem (3.1.1)–(3.1.6)) has a unique solution $u(z,\xi)$. This problem we call the direct problem 3.1.

Now, we consider the case

$$v(z) > v_0/\sin\theta_0, \qquad 0 \le z < \infty, \tag{E}$$

when differential equation (3.1.7) in the two-dimensional reduced problem has the elliptic type everywhere in the half-plane $z \ge 0$, $-\infty < \xi < \infty$. We introduce the positive function $\overset{+}{v}$ by the formula

$$\frac{1}{\overset{+}{v}{}^2(z,\theta_0)} = \frac{\sin^2\theta_0}{v_0^2} - \frac{1}{v^2(z)}, \qquad 0 \le z < \infty.$$

In the elliptic case (E), we obtain the following direct problem: in the half-plane $z > 0$, $-\infty < \xi < \infty$, we are given the equation

$$\frac{\partial^2 u}{\partial z^2} + \eta(z)\,\frac{\partial^2 u}{\partial \xi^2} + \tau(z)\,\frac{\partial u}{\partial z} = 0, \qquad 0 < m < \eta(z) < M < \infty, \tag{3.1.12}$$

where

$$\tau(z) = \frac{\mathrm{d}}{\mathrm{d}z}\{\ln\mu(z)\}, \qquad \eta(z) = \frac{1}{\overset{+}{v}{}^2(z,\theta_0)}.$$

When $z = 0$, $-\infty < \xi < \infty$, the boundary condition for directional derivative is set:

$$\left(\frac{\partial u}{\partial z} - \varkappa\frac{\partial u}{\partial \xi}\right)\bigg|_{z=0} = f(\xi), \tag{3.1.13}$$

where

$$\varkappa = \frac{\delta}{\mu(0)}, \qquad \delta = \cos\theta_0\,\frac{\mu_0}{v_0}, \qquad f(\xi) = -2\varkappa\,\frac{\mathrm{d}\varphi_0}{\mathrm{d}\xi}. \tag{3.1.14}$$

Besides, we set the following condition: $u(z,\xi) \to 0$ uniformly relatively $\xi \in (-\infty,\infty)$ and $z \in [0,\infty)$ when $z \to \infty$ or when $|\xi| \to \infty$ respectively (condition (3.1.15)).

We suppose, also, that $u(z,\xi)$ satisfies the smoothness conditions formulated in the boundary-value problem 2.1, Chapter 2. We need to find the function $u(z,\xi)$ satisfies all these conditions.

We shall call this problem by the direct problem 3.2. As a certain particular case of the direct problem 2.1, the problem 3.2 may have a single solution $u(z,\xi) = u(z,\xi,\theta_0)$.

Remark 3.1.2. In the variables x, z, t, condition (3.1.15) is equivalent to the following condition: for $t = 0$ (and for each other moment t also) the function $u(x, z, t)$ (shift field ξ in the layer) tends to zero for $z \to \infty$ or for $|x| \to \infty$, i. e., when we move off the point of intersection of the front of the incident wave φ_0 with the boundary $z = 0$ uniformly relatively $x \in (-\infty; \infty)$ and $z \in [0, \infty)$ respectively.

3.2. DETERMINATION OF PROPERTIES OF INHOMOGENEOUS LAYER BY THE FORMS OF INCIDENT AND REFLECTED WAVES GIVEN FOR A SINGLE ANGLE θ_0

Inverse problem 3.1 (for the hyperbolic case).

We suppose that for $0 \le \xi < \infty$ and for a certain fixed $\theta_0 \in [0, \pi/2)$ the two functions $\varphi_0(\xi, \theta_0)$ and $\varphi_1(\xi, \theta_0)$ are given. We suppose also that $\varphi_1(\xi, \theta_0)$ is connected with the solution $u(z, \xi, \theta_0)$ of the direct problem 3.1 by condition (3.1.10) with certain unknown functions $v(z)$, $\mu(z)$ $(0 \le z < \infty)$ and unknown numbers v_0 and μ_0; i. e., the forms of incident and reflected waves are given. The number $\delta = \cos\theta_0 \cdot \sigma^0$, is given also, where $\sigma^0 = \mu_0/v_0$. We obtain such information (in the problem with plane waves) if we measure the incident and reflected waves as the function dependent on $t \in (-\infty; \infty)$ in a certain point (x_0, z_0) of the domain $z \le 0$ and set, for example, the numbers θ_0, v_0, and μ_0. These data we shall call the data of the inverse problem 3.1; these data relate to a fixed value of the parameter θ_0.

We need to find: in the general case, for given θ_0, for $x \ge 0$ the function $\sigma(x, \theta_0)$. This function is obtained from the function

$$\sigma(z, \theta_0) = \mu(z)/\bar{v}(z, \theta_0)$$

by the change of variables

$$x = \int_0^z \frac{\mathrm{d}z}{\bar{v}(z, \theta_0)}, \qquad \sigma(x, \theta_0) = \bar{\sigma}(z, \theta_0) \equiv \sigma(x(z, \theta_0), \theta_0).$$

If $\mu(z) \equiv \text{const} = \mu$ is set and the number v_0 is set, then we must find the function $v(z)$ for $0 \le z < \infty$. If the number v_0 and the function $v(z)$ for $0 \le z < \infty$ are given, then we must find the function $\mu(z)$ for $0 \le z < \infty$.

Note that equalities (3.1.8′) are the relations of type (1.4.1). In this case, the value $\cos\theta_0\sigma^0$ plays the role of the number δ, and $\varphi^*(\xi) = -2\varphi_0(\xi)$. Further, from condition (3.1.10), given the functions φ_0, φ_1, we find the function

$u(0,\xi) = u(0,\xi,\theta_0)$. Thus, the inverse problem 3.1 for the hyperbolic case is equivalent to the inverse problem 1.2. Therefore, all the results of Chapter 1: Theorem 1.2.1 and the method of solution of the inverse problem 1.2 are carried over the inverse problem 3.1. Moreover, from Theorem 1.4.1 the following theorem holds immediately:

Theorem 3.2.1. *Denote*

$$B(x) = -\frac{1}{2}\frac{\mathrm{d}}{\mathrm{d}x}\{\ln \sigma^{-1}(x,\theta_0)\}, \quad q(x) = \frac{\mathrm{d}B}{\mathrm{d}x} + B^2. \tag{3.2.1}$$

Suppose that the following conditions hold:

(A) the derivatives $\mathrm{d}^2 v/\mathrm{d}z^2$ *and* $\mathrm{d}^2\mu/\mathrm{d}z^2$ *are continuous for* $z \geq 0$;

(B) $\int_0^\infty x|q|\,\mathrm{d}x < \infty$, $\int_0^\infty B^2\,\mathrm{d}x < \infty$, $\int_0^\infty |B|\,\mathrm{d}x < \infty$,

(C) for fixed θ_0, *the function* $\varphi_0(\xi,\theta_0)$ *and its derivatives with respect to* ξ *up to the fourth order are continuous and absolutely integrable in the line* $-\infty < \xi < \infty$.

Then the inverse problem 3.1 has a unique solution.

Inverse problem 3.2 (the elliptic case).

We suppose that for $-\infty < \xi < \infty$, for a certain fixed $\theta_0 \in (0,\pi/2)$ the functions $\varphi_0(\xi,\theta_0)$ and $\varphi_1(\xi,\theta_0)$ are set, where $\varphi_1(\xi,\theta_0)$ are connected with the solution $u(z,\xi,\theta_0)$ of the direct problem 3.2 by condition (3.1.10) with certain unknown functions $v(z)$, $\mu(z)$ $(0 \leq z < \infty)$ and unknown numbers v_0 and μ_0; i. e., the forms of incident and reflected waves are set. The number $\delta = \cos\theta_0\sigma^0$ is given also.

We need to find: in the general case, for given θ_0, for $x \geq 0$, the function $\overset{+}{\sigma}(x,\theta_0)$, which is obtained from the function

$$\overset{+}{\sigma}(z,\theta_0) = \mu(z)/\overset{+}{v}(z,\theta_0)$$

by the change of variables

$$x = \int_0^z \frac{\mathrm{d}z}{\overset{+}{v}(z)}, \qquad \overset{+}{\sigma}(x,\theta_0) = \overset{+}{\sigma}(z,\theta_0) \equiv \overset{+}{\sigma}(x(z,\theta_0),\theta_0).$$

If $\mu(z) \equiv \text{const} = \mu$ and the numbers μ, v_0 are given, then we find the function $v(z)$ for $0 \leq z < \infty$; if the number v_0 and the function $v(z)$ is set for $0 \leq z < \infty$, then we must find the function $\mu(z)$ for $0 \leq z < \infty$.

Evidently, the inverse problem is equivalent to the inverse problem 2.2. Therefore, all the results of Chapter 2: Theorem 2.2.1 and the method of solution of the inverse problem 2.2 are carried over the inverse problem 3.2. Moreover, from Theorem 2.3.2 follows:

Theorem 3.2.2. *Let the conditions (A)–(C) of Theorem 3.2.1 hold if we replace the function* $\sigma(x, \theta_0)$ *by* $\overset{+}{\sigma}(x, \theta_0)$ *in formulas (3.2.1). Let, also, the condition*

(D) $\tilde{\varphi}_0(k)|_{k=0} = 0$

hold, where $\tilde{\varphi}_0(k)$ *is the Fourier transform of the function* $\varphi_0(\xi) = \varphi_0(\xi, \theta_0)$. *Then the inverse problem 3.2 has a unique solution.*

Thus, in the elliptic case, we succeed in determination of similar characteristics of inhomogeneous layer (half-space) given similar data as in the hyperbolic case.

The condition (B) in Theorems 3.2.1 and 3.2.2 hold, for example, for the functions $v(z)$ and $\mu(z)$ which are constant outside a certain segment $[0, H]$. Physically it corresponds to the transition layer of the variable velocity $v(z)$ with the density $\rho(z)$. As it follows from the conditions (A), (B) of both theorems, the speed $v(z)$ of propagation of SH-waves in the inhomogeneous layer may be not monotone. So, in the general case, given the data $\{\varphi_0(\xi, \theta_0), \varphi_1(\xi, \theta_0)\}$ for one value θ_0, we can determine uniquely only the function $\sigma(x, \theta_0)$ or $\overset{+}{\sigma}(x, \theta_0)$. The functions $v(z)$ and $\rho(z)$ could not be determined. In order to find them, we shall formulate another inverse problem, where the inverse problem data is given wider than in the inverse problems 3.1 and 3.2. First, we consider the hyperbolic case.

3.3. THE METHOD OF RECOVERY OF THE DENSITY AND THE SPEED IN THE INHOMOGENEOUS LAYER AS THE FUNCTIONS OF THE DEPTH GIVEN THE SET OF PLANE WAVES REFLECTED FROM THE LAYER AT VARIOUS ANGLES

Inverse problem 3.3.

Unlike the inverse problem 3.1, where formulas (3.1.7′) are not needed for determination of the functions $\eta(z)$ and $\bar{v}(z, \theta_0)$, here we use essentially the dependence of the function $\bar{v}(z, \theta_0)$ on the parameter θ_0 of the form (3.1.7′) in the inverse problem 3.3, which holds in the problem with plane waves.

We denote by Θ_0 an infinite monotone increasing (or decreasing) sequence of real numbers $\theta_0^{(1)}, \theta_0^{(2)}, \dots, \theta_0^{(n)}, \dots$ such that

1) each θ_0 from the sequence Θ_0 belongs to the interval $[0, \pi/2)$ and satisfies the condition (H);

2) the sequence Θ_0 converges to a certain finite limit $\bar{\theta}_0 \neq 0$.

We suppose that the conditions (A) and (B) of Theorem 3.2.1 hold and the function $\varphi_0(\xi, \theta_0)$ satisfies the condition (C) of this theorem for each fixed $\theta_0 \in \Theta_0$. Let also, for $0 \leq \xi < \infty$ and for each θ_0 from the sequence Θ_0, the forms of incident and reflected waves $\varphi_0(\xi, \theta_0)$ and $\varphi_1(\xi, \theta_0)$ be given. Here the function $\varphi_1(\xi, \theta_0)$ is connected with the solution $u(z, \xi, \theta_0)$ of the direct problem 3.1 by condition (3.1.10) with certain unknown functions $v(z)$ and $\mu(z)$. The numbers v_0 and μ_0 are given also. Thus, data of the inverse problem 3.1 are given for each $\theta_0 \in \Theta_0$.

We need to find the functions $v(z)$ and $\mu(z)$ for $0 \leq z < \infty$.

Theorem 3.3.1. *The inverse problem 3.3 has a unique solution* $\{v(z), \mu(z)\}$.

We shall prove this theorem and give the method for solution of this problem. All this based on the following auxiliary statement.

Lemma 3.3.1. *Denote*

$$\alpha = -\sin\theta_0 \cos\theta_0 / v_0^2, \qquad \bar{v}(x, \theta_0) = \bar{v}(z, \theta_0),$$

where

$$x = \int_0^z \frac{\mathrm{d}z}{\bar{v}(z, \theta_0)}.$$

Then the identity

$$\alpha \bar{v}^2(x, \theta_0) = \frac{\partial}{\partial x}\{\ln \sigma(x, \theta_0)\} \cdot \alpha \int_0^x \bar{v}^2(x, \theta_0)\,\mathrm{d}x + \frac{\partial}{\partial \theta_0}\{\ln \sigma(x, \theta_0)\}. \quad (3.3.1)$$

holds.

Proof. This lemma follows immediately from the definition of $\sigma(x, \theta_0)$, i. e., from the equality

$$\bar{\sigma}(z, \theta_0) \equiv \sigma(x(z, \theta_0), \theta_0) = \sigma(x, \theta_0).$$

Differentiating the equality with respect to θ_0 in a point (z, θ_0), we obtain

$$\frac{\partial \bar{\sigma}(z,\theta_0)}{\partial \theta_0}\Big|_{z,\theta_0} = \frac{\partial \sigma(x,\theta_0)}{\partial x}\Big|_{x=x(z,\theta_0),\theta_0} \cdot \frac{\partial x(z,\theta_0)}{\partial \theta_0}\Big|_{z,\theta_0} + \frac{\partial \sigma(x,\theta_0)}{\partial \theta_0}\Big|_{x=x(z,\theta_0),\theta_0}. \qquad (!)$$

It is easy to show that

$$\frac{\partial x}{\partial \theta_0} = \alpha \int_0^z \bar{v}(z,\theta_0)\,\mathrm{d}z.$$

Changing the variables in this equality as in the formulation of lemma is mentioned, we obtain

$$\frac{\partial x}{\partial \theta_0} = \alpha \int_0^x \bar{v}^2(x,\theta_0)\,\mathrm{d}x.$$

Further, we have

$$\frac{1}{\bar{\sigma}(z,\theta_0)}\,\frac{\partial \bar{\sigma}(z,\theta_0)}{\partial \theta_0} = \alpha \bar{v}^2(z,\theta_0) = \alpha \bar{v}^2(x,\theta_0).$$

Dividing identity (!) by the function $\bar{\sigma}(z,\theta_0) = \sigma(x,\theta_0)$ and substituting the two last formulas, we obtain identity (3.3.1). □

Proof of Theorem 3.3.1. We return now to the inverse problem 3.3. We suppose that the functions $v(z)$ and $\mu(z)$ are unknown. We solve for each value $\theta_0 \in \Theta_0$ the inverse problem 3.1 (for example, using the method of § 1.3). As a result, we obtain (uniquely, by Theorem 3.2.1) the function $\sigma(x,\theta_0)$ for each $\theta_0 \in \Theta_0$ and for $x \geq 0$. Therefore, we can find the partial derivatives

$$\frac{\partial}{\partial x}\{\ln \sigma(x,\bar{\theta}_0)\} \quad \text{and} \quad \frac{\partial}{\partial \theta_0}\{\ln \sigma(x,\theta_0)\}\Big|_{\theta_0=\bar{\theta}_0}.$$

These derivatives we derive in the limit point $\bar{\theta}_0$ for $x \geq 0$. They are derived uniquely also.

We consider now identity (3.3.1) in the limit point $\bar{\theta}_0$:

$$\bar{\alpha}\bar{v}^2(x,\bar{\theta}_0) = \frac{\partial}{\partial x}\{\ln \sigma(x,\bar{\theta}_0)\} \cdot \bar{\alpha} \int_0^x \bar{v}^2(x,\theta_0)\,\mathrm{d}x + b(x), \qquad (3.3.2)$$

where

$$b(x) = \frac{\partial}{\partial\theta_0}\{\ln\sigma(x,\theta_0)\}\Big|_{\theta_0=\bar\theta_0}, \qquad \bar\alpha = -\frac{\sin\bar\theta_0\cos\bar\theta_0}{v_0^2}.$$

As $\bar\theta_0 \neq 0$, then $\bar\alpha \neq 0$. Since, for $x = 0$, we have $b(0) = \bar\alpha\bar v^2(x,\theta_0) \neq 0$, the function $b(x)$ is not equal to zero in a certain interval $[0,\delta)$ as a continuous function. Therefore, equality (3.3.2) is an integral Volterra equation of the second kind

$$y(x) = \int_0^x K(x,\xi)y(\xi)\,\mathrm{d}\xi + b(x)$$

with the continuous (degenerate) kernel

$$K(x,\xi) = \frac{\partial}{\partial x}\{\ln\sigma(x,\bar\theta_0)\}$$

and continuous right–hand side

$$b(x) = \frac{\partial}{\partial\theta_0}\{\ln\sigma(x,\theta_0)\}\Big|_{\theta_0=\bar\theta_0}$$

relative the unknown function $y(x) = \bar\alpha\bar v^2(x,\theta_0)$.

As is known from theory of integral equations (Petrovskii, 1951, p. 56), such equation has a unique solution $y(x)$ in each finite segment $[0,a]$ from the x-axis.

Moreover, the solution $\bar v^2(x,\bar\theta_0)$ of equation (3.3.2) has the explicit form

$$\bar v^2(x,\bar\theta_0) = \frac{1}{\bar\alpha}\Big\{b(x) + \frac{\partial\sigma(x,\bar\theta_0)}{\partial x}\int_0^x \frac{b(x)}{\sigma(x,\bar\theta_0)}\,\mathrm{d}x\Big\}, \tag{3.3.3}$$

which can be shown if we substitute the right-hand side of (3.3.3) into equality (3.3.2) and integrate by parts.

Deriving by formula (3.3.3) the function $\bar v(x,\bar\theta_0)$, we find then the function $\mu(x,\bar\theta_0) = \bar v(x,\bar\theta_0)\sigma(x,\bar\theta_0)$. Then we find the function

$$z = z(x,\bar\theta_0) = \int_0^x \bar v(x,\bar\theta_0)\,\mathrm{d}x$$

and the functions

$$\bar v(z,\bar\theta_0) = \bar v(x,\bar\theta_0), \qquad \mu(z) = \mu(x,\bar\theta_0),$$

where $z = z(x,\bar\theta_0)$. At last, by formula (3.1.7′) and taking into account that $\rho(z) = \mu(z)/v^2(z)$, we recover the desired functions $v(z)$ and $\rho(z)$ for $z \geq 0$. Theorem 3.3.1 is proved. □

3.4. THE ALGORITHM OF NUMERICAL SOLUTION OF THE INVERSE PROBLEM 3.3 (DETERMINATION OF $v(z)$ AND $\rho(z)$ BY THE FORMS $\varphi_0(\xi,\theta_0)$, $\varphi_1(\xi,\theta_0)$ OF INCIDENT AND REFLECTED WAVES FOR THREE ANGLES θ_0)

The formula (3.3.3) is basic for this method.

Following the method stated in § 3.3, the algorithm has the two stages of recovering the functions $v(z)$ and $\rho(z)$.

1. Here we find the function $\sigma(x) = \sigma(x,\theta_0)$ for a fixed θ_0 given the pair of functions $\varphi_0(\xi) = \varphi_0(\xi,\theta_0)$ and $\varphi_1(\xi) = \varphi_1(\xi,\theta_0)$ (i. e., we solve the inverse problem 3.1).

   This inverse problem is solving three times, for three angles: $\theta - \Delta\theta$; θ; $\theta + \Delta\theta$, where $\Delta\theta$ is small. As a result, we obtain three functions

$$\sigma_1(x) = \sigma(x,\theta - \Delta\theta), \quad \sigma_2(x) = \sigma(x,\theta), \quad \sigma_3(x) = \sigma(x,\theta + \Delta\theta).$$

2. We recover the functions $v(z)$ and $\rho(z)$ using the functions $\sigma_1(x), \sigma_2(x), \sigma_3(x)$ and the difference analog of formula (3.3.3).

3.4.1. Determination of $\sigma(x,\theta_0)$ for a fixed θ_0. As the parameter θ_0 in this section is fixed, we denote, for brevity,

$$\sigma(x,\theta_0) = \sigma(x), \qquad u(z,\xi,\theta_0) = u(z,\xi),$$
$$\varphi_0(\xi,\theta_0) = \varphi_0(\xi), \qquad \varphi_1(\xi,\theta_0) = \varphi_1(\xi).$$

For numerical deriving the function $\sigma(x)$, given the functions $\varphi_0(\xi)$ and $\varphi_1(\xi)$, we use the method of Antonenko (1967). This is the method of solution of the inverse one-dimensional dynamic problem of seismic which is based on applying the principle of feedback.

To this end, we pass to the variables

$$x = \int_0^z \frac{\mathrm{d}z}{\bar{v}(z,\theta_0)}, \qquad u(x,\xi) = u(z,\xi)$$

and obtain, instead of the direct problem 3.1, the following problem equivalent to it. The numbers θ_0, $\sigma^0 = \mu_0/v_0$ are given; the functions $\sigma(x)$, for

$x \geq 0$ and $\varphi_0(\xi)$ for $-\infty < \xi < \infty$ are given in the conditions

$$\frac{\partial^2 u}{\partial x^2} + \frac{\mathrm{d}}{\mathrm{d}x}\{\ln \sigma(x)\}\frac{\partial u}{\partial x} = \frac{\partial^2 u}{\partial \xi^2}, \qquad x > 0, \quad -\infty < \xi < \infty, \tag{3.4.1}$$

$$\frac{\partial u}{\partial x}(0,\xi) - \cos\theta_0 \frac{\sigma^0}{\sigma(0)}\frac{\partial u}{\partial \xi}(0,\xi) = -2\cos\theta_0 \frac{\sigma^0}{\sigma(0)}\frac{\mathrm{d}\varphi_0}{\mathrm{d}\xi}, \tag{3.4.2}$$

$$u(x,\xi)\big|_{\xi<0} \equiv 0, \qquad x \geq 0. \tag{3.4.3}$$

We need to find the function $u(x,\xi)$ satisfying conditions (3.4.1)–(3.4.3) and the function $\varphi_1(\xi)$ from the condition

$$u(0,\xi) = \varphi_0(\xi) + \varphi_1(\xi). \tag{3.4.4}$$

Problem (3.4.1)–(3.4.3) is differed from the problem considered by Antonenko (1967) only by the condition (3.4.2): it sets not the normal derivative $\partial u/\partial x(0,\xi)$ but the directional derivative. All stated in this section is the transfer of the method of Antonenko (1967) on the problem (3.4.1)–(3.4.3).

Parallel with problem (3.4.1)–(3.4.4), we consider the difference problem. For this purpose, we construct the square net in the half-plane $x \geq 0$, $-\infty < \xi < \infty$, with the step h consisted from the points with the coordinates

$$x = ih \quad (i = 0,1,\ldots), \qquad \xi = kh \quad (k = \ldots,-1,0,1,\ldots).$$

We change the derivatives entering conditions (3.4.1)–(3.4.3) by the difference relations

$$\frac{\partial u}{\partial x} \sim \frac{u_{i+1}^k - u_{i-1}^k}{2h}, \qquad \frac{\partial u}{\partial \xi} \sim \frac{u_i^{k+1} - u_i^{k-1}}{2h}, \qquad \frac{\mathrm{d}\sigma(x)}{\mathrm{d}x} \sim \frac{\sigma_i - \sigma_{i-1}}{h},$$
$$\frac{\partial^2 u}{\partial x^2} = \frac{u_{i-1}^k - 2u_i^k + u_{i-1}^k}{h^2}, \qquad \frac{\partial^2 u}{\partial \xi^2} \sim \frac{u_i^{k+1} - 2u_i^k + u_i^{k-1}}{h^2}.$$

The value of the net function u^h in the point (ih, kh) we denote by u_i^k; $\sigma_i = \sigma(ih)$. As a result, we obtain the difference equation

$$u_i^{k+1} + u_i^{k-1} = A_i u_{i+1}^k + B_i u_{i-1}^k, \tag{3.4.5}$$

where

$$B_i = 1 - \frac{\sigma_i - \sigma_{i-1}}{2\sigma_i}, \quad A_i = 2 - B_i, \quad i = 1,2,\ldots; \; k = \ldots,-1,0,1,\ldots;$$

the difference boundary condition

$$u_0^k = (u_1^k + cu_0^{k-1} + 2\,\mathrm{ch}\, f_k)/(1+c), \tag{3.4.6}$$

where

$$c = \cos\theta_0 \, \frac{\sigma^0}{\sigma(0)}, \quad f_k = \frac{\mathrm{d}\varphi_0}{\mathrm{d}\xi}(kh), \quad k = 0, 1, 2, \ldots$$

($f_k = 0$ for $k < 0$); and the difference initial condition

$$u_i^k \equiv 0 \quad \text{for} \quad k < i. \tag{3.4.7}$$

Besides, we have the condition

$$u_0^k = \varphi_0^k + \varphi_1^k, \tag{3.4.8}$$

where

$$\varphi_0^k = \varphi_0(kh), \qquad \varphi_1^k \sim \varphi_1(kh), \qquad k = 0, 1, \ldots$$

($\varphi_0^k = 0$ for $k < 0$).

We shall formulate the direct problem as follows. The numbers h (the net step), θ_0, and σ^0 are given. The net functions f_k, φ_0^k are set from conditions (3.4.6), (3.4.8) for $k = 0, 1, \ldots$; σ_i are set for $i = 0, 1, \ldots$. So, the net functions A_i and B_i are set from equation (3.4.5) for $i = 1, 2, \ldots$ and the number c is given also. We need to find the net function φ_1^k from condition (3.4.8), where the function u_i^k satisfies conditions (3.4.5)–(3.4.7).

Firstly, we should define the values u_i^k in the boundary net points, for $i = 0$. From expression (3.4.6) and taking into account condition (3.4.7), we find the value u_0^0 for $k = 0$, and from equation (3.4.5), for $k = i - 1$, we obtain

$$u_i^i = B_i u_{i-1}^{i-1}, \qquad i = 1, 2, \ldots.$$

This gives us the values u_i^k in the first diagonal $i = k$. Then, from (3.4.6) for $k = 1$ we find the value u_0^1, and from (3.4.5) for $k = i$, we have

$$u_i^{i+1} = B_i u_{i-1}^i, \qquad i = 1, 2, \ldots.$$

As a result we define the values u_i^k in the second diagonal $k = i + 1$. Extending this recurrent process, we define the values u_0^k; then, from condition (3.4.8), we find the desired net function φ_1^k for $k = 0, 1, 2, \ldots$.

If the net function σ_i is set in a finite number of points for $i = 0, 1, \ldots, K$, and the functions f_k and φ_0^k are set for $k = 0, 1, \ldots, N$, where $N = 2(K-1)$, then the function φ_1^k can be determined for $k = 0, 1, \ldots, N$ also.

We formulate now the difference inverse problem correspondent to the inverse problem 3.1.

Let, for $k = 0, 1, \ldots, N = 2(K - 1)$, a net function φ_1^k be given, which solves direct problem (3.4.5)–(3.4.8). The net function σ_i and, therefore, the coefficients A_i, B_i are unknown for $i = 1, 2, \ldots$. Besides, let the numbers c, σ_0, h be given and the net functions φ_0^k and $f(k)$, $k = 0, 1, \ldots, N$ be derived following conditions (3.4.6), (3.4.8).

We need to find the net function σ_i for $i = 1, 2, \ldots, K-1$.

First, we find u_0^k from condition (3.4.8) and then, the function u_1^k from condition (3.4.6).

Using the net functions u_0^k, u_1^k and the numbers h, c we find the net function B_i, $i = 1, 2, \ldots, K-1$ following the method of Antonenko (1967). However, in our case we have one distinction. The formula (9′) from the paper of Antonenko (1967) is the basis of application of feedback and it represents the relation between the error δ_n in the recovered value B_n and the error Δ_n in the value u_0^{2n} caused by δ_n. Instead of this formula we use the dependence

$$\delta_1 = \frac{(1+c)\Delta_1}{u_0^1}, \qquad \delta_n = \frac{(1+c)\Delta_n}{u_{n-1}^n \prod_1^{n-1} A_i}, \qquad n \geq 2. \tag{3.4.9}$$

Here Π is the product,

$$\Delta_n = \tilde{u}_0^{2n} - u_0^{2n}, \qquad \delta_n = \bar{B}_n - B_n, \qquad A_i = 2 - B_i, \quad i = 1, 2, \ldots, n-1,$$

where u_0^{2n} is the given value u_i^k in the point $(0, 2n)$; $\tilde{u}_0^{2n}$ is recovered by $B_1, \ldots, B_{n-1}, \bar{B}_n$ value u_i^k in the point $(0, 2n)$. The values $B_1, \ldots, B_{n-1}$ are the values of the net function B_i being found exactly; i. e., this is the solution of the direct problem $\tilde{u}_0^2, \ldots, \tilde{u}_0^{2n-2}$, constructed by these values which differs from given $u_0^1, \ldots, u_0^{2n-2}$ less than $\varepsilon > 0$ which is given:

$$|\tilde{u}_0^k - u_0^k| \leq \varepsilon, \qquad k = 2, 4, \ldots, 2n-2; \tag{3.4.10}$$

$\bar{B}_n$ is the value of B_n recovered by the values $B_1, \ldots, B_{n-1}$ and u_0^{2n-1} as a result of recurrent determination of $u_1^{2n-1}, u_2^{2n-2}, \ldots, u_n^n$ using the formula

$$B_n = u_n^n / u_{n-1}^{n-1}. \tag{3.4.11}$$

Dependence (3.4.9) is deduced analogously as formula (9′) in Antonenko (1967) and is applied analogously. Derivation of the values u_i^k when we define the next (not revised by the feedback) value B_n $(n = 1, 2, \ldots, K-1)$ is gone by the contour Γ_{2n-1} from the system $\Gamma_1, \Gamma_2, \ldots, \Gamma_{N-1}$. The system of contours is shown in Figure 3.2 by the continuous lines.

In this case we move from the point $(0, 2n-1)$ to the point (n, n) where the value B_n is found by formula (3.4.11). Derivation of $\tilde{u}_i^k$ when we revise B_n using the feedback is going along the contour Γ_{2n} from the system

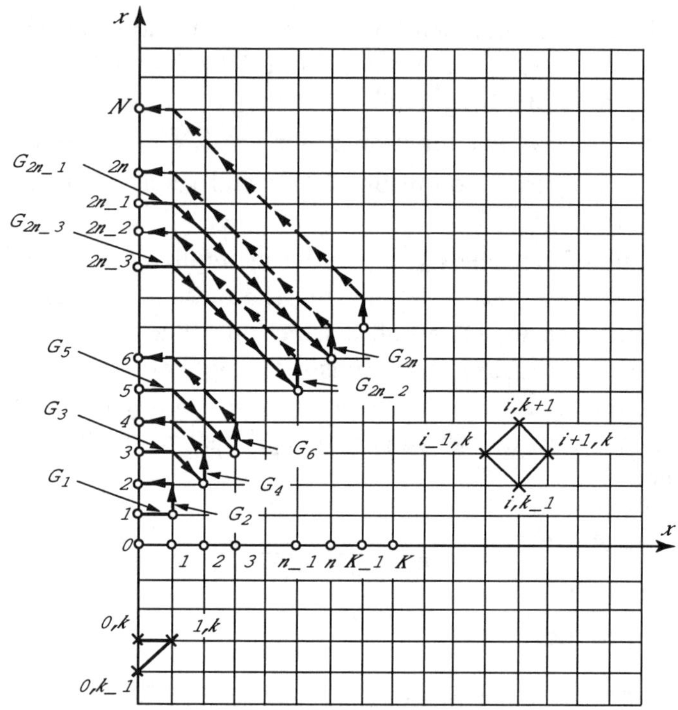

Figure 3.2. To the solution of the difference inverse problem

of contours $\Gamma_2, \Gamma_4, \ldots, \Gamma_N$ shown in Figure 3.2 by the dotted line. In this case, we move from the point (n, n) to the point $(0, 2n)$, where we define $\tilde{u}_0^{2n}$. Then we compare by formula (3.4.10) and revise the obtained value B_n by the formula $B_n \to B_n - \delta_n$. The process of revising we continue up to the moment when the inequality $|\tilde{u}_0^{2n} - u_0^{2n}| < \varepsilon$ holds. Then we pass to the next contour Γ_{2n+1} and define the next nonrevised value B_{n+1} and so on.

Using the recovered values B_i we define then the desired net function σ_i by the recurrent formula

$$\sigma_i = \sigma_{i-1}/(2B_i - 1), \qquad i = 1, 2, \ldots, K-1 \qquad (\sigma_0 \text{ is given}).$$

3.4.2. Numerical determination of the functions $v(z)$ and $g(z)$ given $\{\varphi_0(\xi, \theta_0), \varphi_1(\xi, \theta_0)\}$ for three various angles θ_0. We take three close values of the parameter θ_0:

$$\theta_0^{(1)} = \theta - \Delta\theta, \qquad \theta_0^{(2)} = \theta, \qquad \theta_0^{(3)} = \theta + \Delta\theta,$$

where $\Delta\theta$ is small.

For each $\theta_0^{(j)}$ $(j = 1, 2, 3)$, we set the data of the difference inverse problem 3.1, i. e. the net functions φ_0^k, φ_1^k (the forms of incident and reflected waves) for $k = 0, 1, \ldots, N = 2(K-1)$ and the number $\sigma^0 = \rho_0 v_0$. Besides, in order to know the value

$$\sigma_0 = \sigma(0, \theta_0^{(j)}) = \mu(0)/\bar{v}(0, \theta_0^{(j)})$$

for each $\theta_0^{(j)}$ $(j = 1, 2, 3)$, the set numbers $v(0)$ and $\rho(0)$ (in the inverse problem 3.3, § 3.3 these numbers were not set).

Suppose that following the method of solving the inverse problem (Subsection 3.4.1), for fixed θ_0, which is equal to $\theta_0^{(j)}$ $(j = 1, 2, 3)$, the three net functions

$$\sigma_i^{(j)} = \sigma_j(x_i) = \sigma(x_i, \theta_0^{(j)}) = \sigma(x_i, \theta + (j-2)\Delta\theta), \qquad j = 1, 2, 3$$

are obtained. Here $x_i = ih$, $i = 0, 1, \ldots, K-1$.

Note that the form φ_0^k of incident impulse for various θ_0 may differ essentially (since each function $\sigma_i^{(j)}$ $(j = 1, 2, 3)$ is defined independently).

Change the variables entering formula (3.3.3) by the difference relations

$$\frac{\partial \sigma(x, \bar{\theta}_0)}{\partial x} \sim \frac{\sigma(x+h, \theta) - \sigma(x-h, \theta)}{2h} = \frac{\sigma_{i+1}^{(2)} - \sigma_{i-1}^{(2)}}{2h},$$

$$\frac{\partial \sigma}{\partial \theta_0}(x, \bar{\theta}_0) \sim \frac{\sigma(x, \theta + \Delta\theta) - \sigma(x, \theta - \Delta\theta)}{2\Delta\theta} = \frac{\sigma_i^{(3)} - \sigma_i^{(1)}}{2\Delta\theta},$$

so that

$$b(x) = \frac{\sigma_i^{(3)} - \sigma_i^{(1)}}{2\Delta\theta\sigma_i^{(2)}}.$$

Here x runs through the points of uniform net with the step h:

$$x = x_i = ih, \qquad i = 1, 2, \ldots, K-2.$$

As a result we obtain the difference analog of formula (3.3.3) for derivation of the function $\bar{v}^2(x, \theta)$:

$$\bar{v}^2(x, \theta) \approx y_i = \frac{1}{\alpha}\left\{\frac{\sigma_i^{(3)} - \sigma_i^{(1)}}{2\Delta\theta\sigma_i^{(2)}} + \frac{\sigma_{i+1}^{(2)} - \sigma_{i-1}^{(2)}}{2h}\int_0^{x_i} \frac{b(x)}{\sigma(x, \theta)}\,\mathrm{d}x\right\}, \qquad (3.4.12)$$

where

$$\alpha = -\sin\theta\cos\theta/v_0^2, \qquad i = 1, 2, \ldots, K-2.$$

Further we define the functions $v(x,\theta)$ and $\rho(x,\theta)$ by the formulas

$$v(x,\theta) = \Big(\frac{1}{\bar{v}^2(x,\theta)} + \frac{\sin^2\theta}{v_0^2}\Big)^{-1/2}, \qquad \rho(x,\theta) = \frac{\sigma(x,\theta)\bar{v}(x,\theta)}{v^2(x,\theta)}$$

where $x = x_i$ $(i = 0, 1, \dots, K-2)$ are the points of the uniform net with the step h. Changing the variables

$$z_i = z(x_i,\theta) = \int_0^{x_i} \bar{v}^2(x,\theta)\,\mathrm{d}x, \qquad v(z_i) = v(x_i,\theta), \quad \rho(z_i) = \rho(x_i,\theta),$$

we obtain the values of the functions $v(z)$ and $\rho(z)$ defined on the nonuniform net $z_1, z_2, \dots, z_{K-2}$ in the z-axis. Finally, we define (using linear or piecewise-cube interpolation) the values $v(z)$ and $\rho(z)$ on the uniform net z with the step Δz.

Remark 3.4.1. Changing the derivative $\partial\sigma/\partial\theta_0$ in (3.3.3) by another difference relation we obtain the formula for derivation of $v(x)$ and $\rho(z)$ given the data $\{\varphi_0(\xi,\theta_0),\ \varphi_1(\xi,\theta_0)\}$ for two values of θ_0 (not for three values).

3.5. DERIVATION OF THE SPEED $v(z)$ AND THE DENSITY $\rho(z)$ IN THE NUMERICAL EXPERIMENTS

In order to clarify the efficiency of the algorithm described in § 3.4 for recovery of the speed $v(z)$ of cross waves propagation and the density $\rho(z)$ in the inhomogeneous layer (proximity and the depth of their propagation, the time of derivation etc.), the algorithm was realized on some control examples.

3.5.1. Recovery of the functions $\sigma(x,\theta_0)$, $v(z)$ and $\rho(z)$ without noise in the initial data. The medium model with essentially nonmonotone change of the speed $v(z)$ and the density $\rho(z)$ was considered. One example with the parameters $v_0 = 1.5$, $\rho_0 = 2$ and the functions

$$v(z) = \begin{cases} 2.5 + 0.5(z - \cos 5\pi z), & 0 \le z \le 1, \\ v(1), & 1 < z \le 10 \end{cases} \qquad \text{(in km/sec)},$$

$$\rho(z) = \begin{cases} 2.4 + 0.2(z - \cos 5\pi z), & 0 \le z \le 1, \\ \rho(1), & 1 < z \le 10 \end{cases} \qquad \text{(in g/sm}^3\text{)}$$

in the inhomogeneous layer of the depth 10km is shown in Figure 3.3. Given these values $v(z)$ and $\rho(z)$ on the net with the step $dz = 0.01$ km and the value θ_0, the values σ_i of the function $\sigma(x, \theta_0)$ were calculated on the uniform net with the step $h = 0.005$sec. Then, the form of incident impulse was taken (see Figure 3.4)

$$\varphi_0(\xi) = e^{-k\xi} \sin 2\pi\omega\xi \qquad (k > 0).$$

After, the theoretical seismogram φ_1^k was derived. This is the solution of direct problem (3.4.5)–(3.4.8) with the step h. Then, using the method of § 3.4 with the parameters h and $\varepsilon = 10^{-5}$, the function σ_i was recovered in the net with the same step h. For each value θ_0 equal to $0, \pi/36, \pi/18, \pi/12, \pi/8$, each of three functions $\varphi_0(\xi)$ from Figure 3.4 was taken.

In all the cases the function σ_i was recovered to within $10^{-3} \div 10^{-5}$ in 587–287 points. Note that, taking into account formula (3.1.7′), the number of points correspondent to the interval with respect to z decreased when θ_0 increases.

Further, given three theoretical seismograms φ_1^k calculated for the values

$$\theta_0^{(1)} = \theta - \Delta\theta, \qquad \theta_0^{(2)} = \theta, \qquad \theta_0^{(3)} = \theta + \Delta\theta,$$

the functions $v(z)$ and $\rho(z)$ were recovered. In this case, the form φ_0^k of incident impulse was taken similarly for each $\theta_0^{(j)}$ $(j = 1, 2, 3)$. The problem was solved for $\theta = \pi/36; \pi/18; \pi/12; \pi/8$ for each of three impulses φ_0 in Figure 3.4 with $\Delta\theta = \pi/1440$. The x-interval, where three functions φ_1^k are set, depends on θ and changes from 5.75sec (1150 points), for $\theta = \pi/36$, to 2.78 sec (556 points) for $\theta_0 = \pi/8$.

In all the examples the functions $v(z)$ and $\rho(z)$ are recovered to the depth 10 km in 1000 points with the step $\Delta z = dz = 10$m. The maximal relative errors in deriving $\rho(z)$ and $v(z)$ (these functions are compared with the initial functions that we take) are not greater than 3–5 %; the general accuracy is much higher.

The problem of recovery $v(z)$ and $\rho(z)$ was solved with similar accuracy for other medium models. One example is pointed in Figure 3.8.

When θ increases, the functions $v(z)$ and $\rho(z)$ are recovered on the same depth more rapidly and with less errors, since for greater θ the number of x-points is less.

3.5.2. Recovery of the functions $\sigma(x, \theta_0)$, $v(z)$ and $\rho(z)$ with the noise being in the initial data. In order to estimate noise stability of the algorithm described above, the random disturbance was taken in the

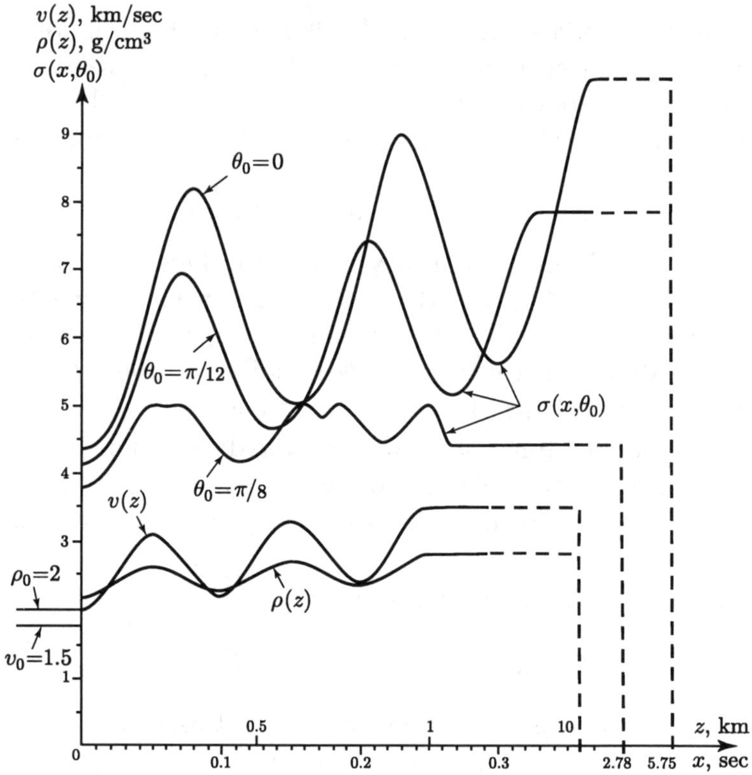

Figure 3.3. Example of the medium model. Recovery of the functions $\sigma(x,\theta_0)$, $v(z)$, and $\rho(z)$:
Three curves $\sigma = \sigma(x,\theta_0)|_{\theta_0=\text{const}}$ constructed (above) for $\theta_0 = 0; \pi/12; \pi/8$ relate to one medium model $\{v(z), \rho(z), v_0, \rho_0\}$ shown below. The functions $\sigma(x,\theta_0)$, $v(z)$, and $\rho(z)$, which are set and recovered by the exact data $\{\varphi_0(\xi,\theta_0), \varphi_1(\xi,\theta_0)\}$ (if the noise is absent), coincide to within the scale of this picture

theoretical seismogram φ_1^k uniformly distributed in the segment $[-\delta,\delta]$, $\delta = nM/100$. Here M is the maximum of absolute value of the function φ_1^k (the useful signal), n is integer (i. e., the error is $n\%$). The form of reflected wave $\tilde{\varphi}_1^k$ with the random error was set as the initial data for recover the function $\sigma(x,\theta_0)$. This inverse problem was solved for the medium model in Figure 3.3 and for the impulses $\varphi_0(\xi)$ and angles θ_0 mentioned in Subsection 3.5.1 with similar parameters $h = 0.005$ sec and $\varepsilon = 10^{-5}$. The level of random noise was varied from 1% to 30%. The algorithm was modified: the function $\tilde{\varphi}_1^k$ was multiply smoothed by the filter of high frequences of the form

$$f_k = (q_{k-1} + 2q_k + q_{k+1})/4.$$

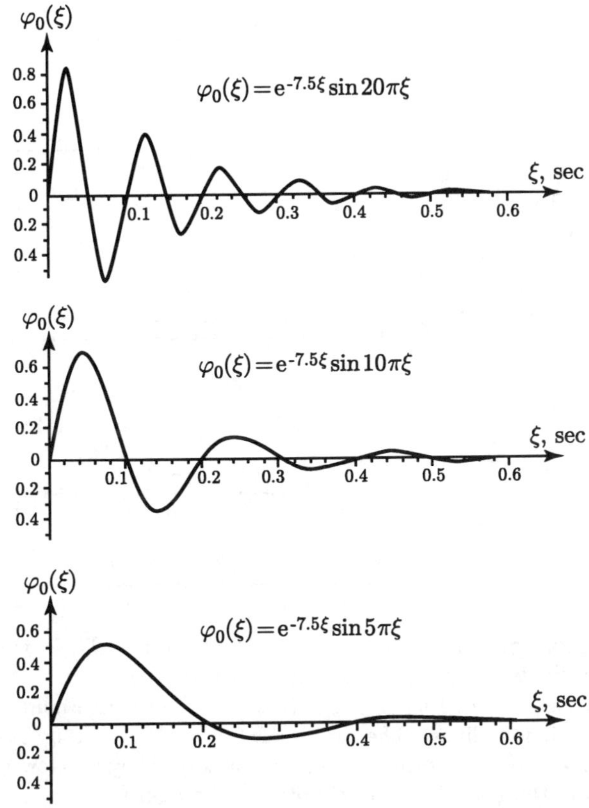

Figure 3.4. The forms of incident impulses $\varphi_0 = \varphi_0(\xi)$ set in the control examples for numerical recovery of the functions $v(z)$, $\rho(z)$, and $\sigma(x, \theta_0)$: In particular, for each of these three impulses, all the functions $v(z)$, $\rho(z)$, and $\sigma(x, \theta_0)$ were recovered. These functions are shown in Figures 3.3 and 3.8

The greater is the noise (n), the greater is the number of filtrations. By the function $\tilde{\varphi}_1^k$ smoothed above we recover the desired function σ_i (that is σ_i^B) using the methods that we apply in the case when the noise is absent.

In Figure 3.5 we see the function σ_i (given) and the function σ_i^B (recovered) for the values

$$\theta_0 = \pi/8, \pi/12, 0, \qquad \varphi_0(\xi) = e^{-7.5\xi} \sin 10\pi\xi$$

and for the various level noise. The following dependence

$$\|\delta\sigma\| = \psi(\|\delta\varphi_1\|)$$

is sufficiently complete characteristic of algorithm stability with respect to the random noise.

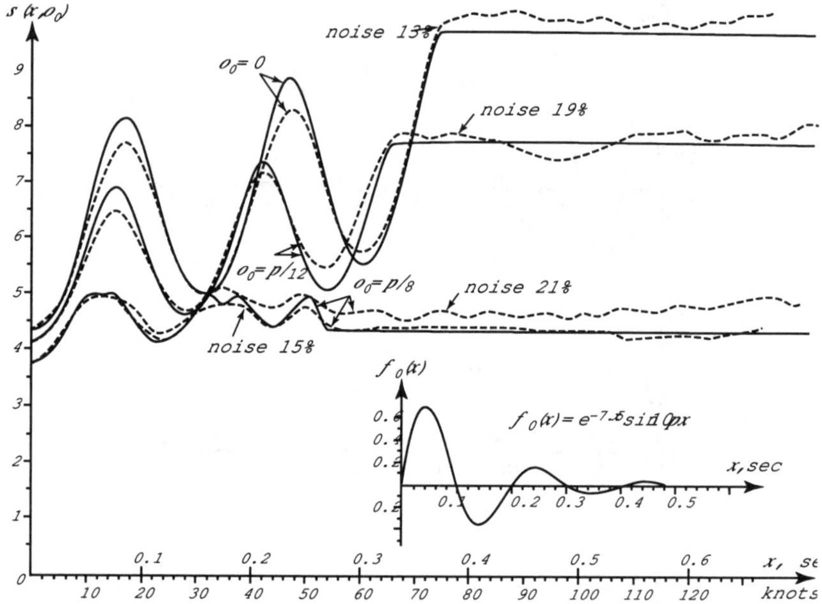

Figure 3.5. Examples of recovery of the function $\sigma(x, \theta_0)$ in the case when the form of reflected wave φ_1^k is given with the error:
The given curves $\sigma = \sigma(x, \theta_0)|_{\theta_0=\text{const}}$ (that are shown in Figure 3.3) are designated by continuous lines. The recovered functions are designated by dotted lines. In all the cases the form of the incident wave shown below was set. The values of θ_0 and the noise level are denoted in the figure

Here $\|\delta\varphi_1\| = n$ (the norm of the noise $\delta\varphi_1^k = \tilde{\varphi}_1^k - \varphi_1^k$ in the initial data), $\|\delta\sigma\|$ is the maximal value of the modulus of the relative error

$$\delta\sigma_i = (\sigma_i^B - \sigma_i)/\sigma_i$$

in % in the whole interval where σ_i is recovering (the norm of the error σ_i). Such dependence, for various θ_0 and $\varphi_0(\xi)$, is shown in Figure 3.6. We see that the accuracy in determination of σ_i increases with increasing θ_0.

The functions $v(z)$ and $\rho(z)$ are recovered by theoretical seismograms φ_1^k set for the number of values

$$\theta_0^{(j)} = \theta + j\Delta\theta, \quad j = -J, \dots, -1, 0, 1, \dots, J; \quad \Delta\theta > 0, \quad J > 1$$

of the parameter θ greater than 3. This is the case when the noise holds. Each of the functions φ_1^k is set with the random error n %. The functions σ_i^B defined by the method described above for each $\theta_0^{(j)}$ are smoothing both with respect to x and θ_0. Formula (3.4.12) is applied then to the three functions $\tilde{\sigma}_i^B$ obtained as the result of such smoothing.

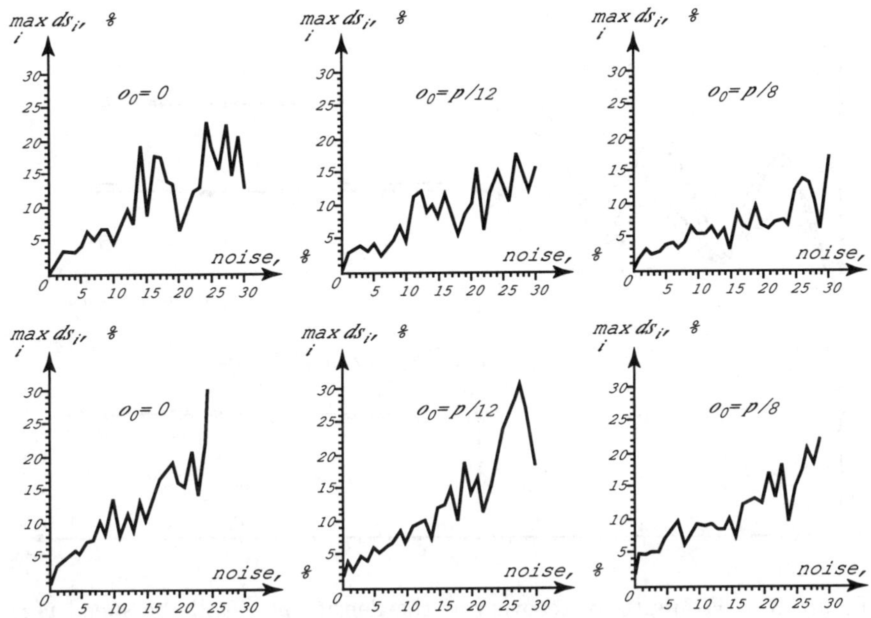

Figure 3.6. Stability of the algorithm of recovery of the function $\sigma(x, \theta_0)$ with respect to the random noise in the initial data:
Each of six graphs shows dependence of max $\delta\sigma_i = \psi(n)$, where $\delta\sigma_i$ is the relative error in deriving $\sigma(x_i, \theta_0)$ in 150 points in the x-axis with the step $h = 0.005$ sec; n is the error level in %, constructed for a fixed θ_0 and fixed form $\varphi_0(\xi)$ of the incident impulse. The three graphs above relate to $\varphi_0(\xi) = e^{-7.5\xi} \sin 10\pi$. The three below graphs relate to $\varphi_0(\xi) = e^{-7.5\xi} \sin 20\pi\xi$. The value θ_0 is shown in the figure. The exact curves $\sigma = \sigma(x, \theta_0)$ are just similar as in Figure 3.3

The calculations show that

1) the number of values θ_0, for which $\tilde{\varphi}_1^k$ is set, is the important parameter which defines the accuracy of recovery of $v(z)$ and $\rho(z)$. The accuracy increases when J increases;

2) this accuracy increases when θ increases. One of the causes of this effect is that the number of points in x-axis where we should derive $\tilde{\sigma}_i^B$ decreases when θ increases.

In Figure 3.7 we have the example of recovery of $v(z)$ and $\rho(z)$ on the depth 2.5 km.

Remark 3.5.1. The formulation of uniqueness theorem for recovery of $v(z)$ and $\rho(z)$ in the elliptic case of the problem of scattering the plane

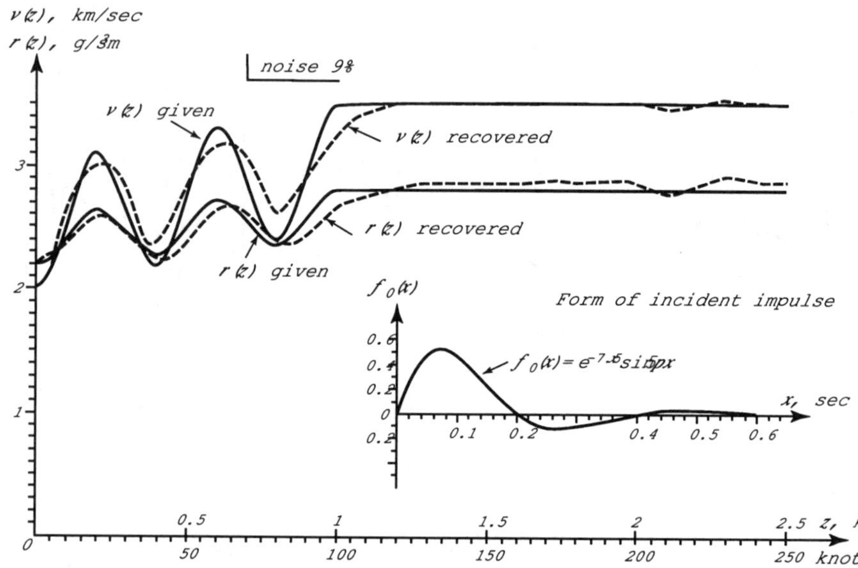

Figure 3.7. Examples of recovery of the density $\rho(z)$ and the speed $v(z)$ in the case when the form of reflected wave φ_1^k is given with the random noise:

The functions $v(z)$ and $\rho(z)$, which are given, are pictured by the continuous line. Their recovery in the case of 9% noise is shown by the dotted line. The form of reflected wave $\tilde{\varphi}_1^k$ is set for $\theta_0 = \theta + j\Delta\theta$, where $j = -15, -14, \dots, -1; 0, 1, \dots, 15$; $\theta = \pi/8$; $\Delta\theta = \pi/1440$

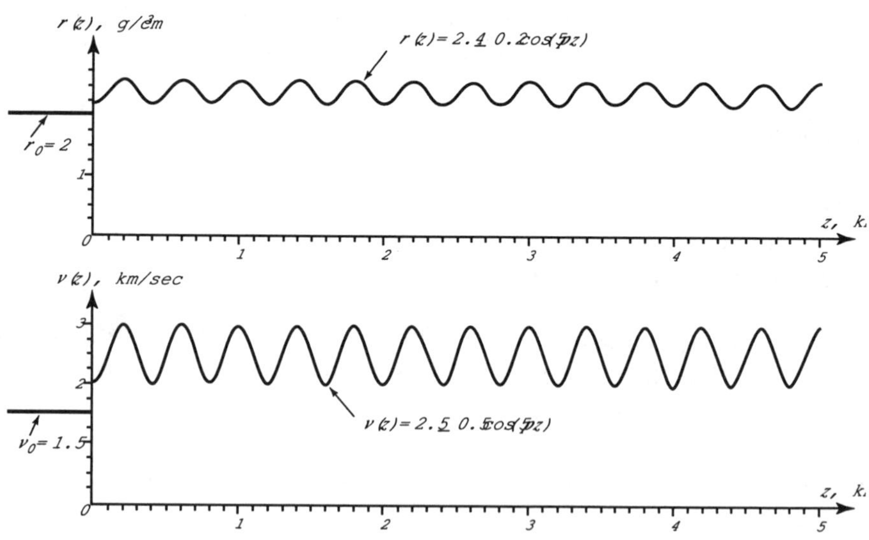

Figure 3.8. Another example of the medium model where the functions $v(z)$ and $\rho(z)$ were recovered:

The functions $v(z)$ and $\rho(z)$ that were given and that were recovered (using the exact data) are coincide to within the scale of this picture waves from the inhomogeneous half-space (transition layer) is analogous to those for the inverse problem 3.3. The role of the inverse problem 3.1 and the formula (3.3.3) play the problem 3.2 and the formula

$$\overset{+}{v}(x,\bar{\theta}_0) = \frac{1}{\overset{+}{\alpha}}\Big\{b(x) + \frac{\partial \overset{+}{\sigma}(x,\bar{\theta}_0)}{\partial x}\int_0^x \frac{b(x)}{\overset{+}{\sigma}(x,\bar{\theta}_0)}\Big\}, \tag{3.5.1}$$

where

$$b(x) = \frac{\partial}{\partial x}\{\ln \overset{+}{\sigma}(x,\theta_0)\}|_{\theta_0=\bar{\theta}_0}, \qquad \overset{+}{\alpha} = \frac{\sin\bar{\theta}_0\cos\bar{\theta}_0}{v_0^2}$$

and functions $\overset{+}{v}(x,\theta_0)$ and $\overset{+}{\sigma}(x,\theta_0)$ are obtained from functions

$$\frac{1}{\overset{+}{v}(z,\theta_0)} = \frac{\sin^2\theta_0}{v_0^2} - \frac{1}{v^2(z)}, \qquad \overset{+}{\sigma}(z,\theta_0) = \frac{\mu(z)}{\overset{+}{v}(z,\theta_0)}$$

by the change of variables

$$x = \int_0^z \frac{dz}{\overset{+}{v}(z,\theta_0)}, \qquad \overset{+}{v}(x,\theta_0) = \overset{+}{v}(z,\theta_0), \qquad \overset{+}{\sigma}(x,\theta_0) = \overset{+}{\sigma}(z,\theta_0).$$

Remark 3.5.2. All the discussions, theorems and formulas of this chapter are transformed easily on the acoustic case. So, we may formulate the analogous inverse problems on the determining the functions $\sigma(x,\theta_0)$, $\overset{+}{\sigma}(x,\theta_0)$ and the speed $v(z)$ of longitudinal waves and the density $\rho(z)$ in the liquid inhomogeneous layer. Instead of equation (3.1.7), in this case, we take the equation for the vertical component U_z of the shift vector $\vec{U}$ obtained by Nikolskii (1965):

$$\frac{\partial}{\partial z}\Big\{\rho(z)v^2(z)\Big[\frac{1}{v^2(z)} - \frac{\sin^2\theta_0}{v_0^2}\Big]\frac{\partial U_z(z,\xi)}{\partial z}\Big\} = \rho(z)\,\frac{\partial^2 U_z(z,\xi)}{\partial \xi^2}.$$

Chapter 4.

Inverse problems for finite string with the condition of directional derivative in one end

In this chapter we consider the coefficient inverse problems for the hyperbolic equation considered for the vibration of the finite string

$$\frac{\partial u}{\partial z}\Big\{\mu(z)\frac{\partial u}{\partial z}\Big\} = \eta(z)\frac{\partial^2 u}{\partial \xi^2}, \quad 0 < z < H, \quad -\infty < \xi < \infty, \tag{4.0.1}$$

given certain data on solution of the (direct) boundary-value problem. On the one end of the string ($z = 0$) we are given the condition of directional derivative

$$\Big(\frac{\partial u}{\partial z} - \varkappa\frac{\partial u}{\partial \xi}\Big)\Big|_{z=0} = f(\xi), \tag{4.0.2}$$

and on the other end ($z = H$) — the condition of the free end. Such data is

either vibrations $u(H, \xi)$ of the free end and the right-hand side of condition (4.0.2),

or the values $u(z, \xi)$ and $\partial u/\partial z$ on the boundary $z = 0$.

The analogous problems are considered for the string with the fixed end $z = H$.

The inverse problems for the string in the case when the boundary conditions

$$\left(\beta\frac{\partial u}{\partial z}+\alpha u\right)\Big|_{z=a_i}=f_i(t),\qquad i=1,2;$$
$$-\infty<\alpha,\beta<\infty;\qquad a_1=0,\qquad a_2=l>0,$$

are set on the both ends of the string are considered by Krein (1951a,b; 1952, 1953, 1954), Parijskii (1969), Gerver (1970). These researches are directly connected with the inverse Sturm—Liouville problem. The boundary conditions of such form generate the self-adjoint differential operator with the real constant coefficients in the boundary conditions and the classical (regular) Sturm—Liouville problem.

In the problem that we consider, the correspondent one-dimensional operator is not self-adjoint and is not Sturm—Liouville operator. The reason is that we have the condition with the directional derivative on the one end. More exactly, we obtain a one-dimensional differential system with a complex variable coefficient in one of the boundary conditions. So we connect this system with a certain set of not self-adjoint operators. Nevertheless, we reduce our inverse problems to the known problem of determining the Sturm—Liouville equation given the two spectra. This problem has the self-adjoint operator (see Chudov, 1949; Krein, 1951; Marchenko, 1952; Gasymov and Levitan, 1964). Namely: first, we consider the two self-adjoint regular Sturm—Liouville problems, connected with our not self-adjoint differential system. Then, using the form of solution of the direct problem and the data of the inverse problems, we determine the spectra of these two problems. This method allows us to solve the inverse problem and to establish the uniqueness theorem.

4.1. FORMULATION OF THE DIRECT PROBLEM

Boundary-value (direct) problem 4.1. Suppose that in the domain $0<z<H$, $-\infty<\xi<\infty$, we are given the equation

$$\frac{\partial}{\partial z}\left\{\mu(z)\frac{\partial u}{\partial z}\right\}=\eta(z)\frac{\partial^2 u}{\partial \xi^2}\quad(\eta(z)>0,\ \mu(z)>0);\tag{4.1.1}$$

the boundary conditions

$$\left(\frac{\partial u}{\partial z}-\varkappa\frac{\partial u}{\partial \xi}\right)\Big|_{z=0}=f(\xi),\tag{4.1.2}$$

$$\frac{\partial u}{\partial z}\Big|_{z=H}=0,\qquad -\infty<\xi<\infty;\tag{4.1.3}$$

and the initial condition

$$u(z,\xi)|_{\xi<0} \equiv 0, \qquad 0 \le z \le H. \tag{4.1.4}$$

Suppose, also, that the following conditions hold:

the function $u(z,\xi)$ and its derivative u_ξ are continuous in the strip $0 \le z \le H$, $-\infty < \xi < \infty$ and the derivatives u_{zz}, $u_{\xi\xi}$, $u_{z\xi}$ are continuous in the strip $0 \le z \le H$, $\xi \ge 0$;

the functions $\eta(z)$, $\mu(z)$, and $(\mathrm{d}\mu)/(\mathrm{d}z)$ are continuous in the segment $0 \le z \le H$;

the number $\varkappa$ is not negative: $\varkappa \ge 0$.

We need to find the function $u(z,\xi)$ defined in the strip $0 \le z \le H$, $-\infty < \xi < \infty$ and satisfying all the conditions mentioned above.

This problem we shall call by the boundary (direct) problem 4.1. Unlike the inverse problems considered below (§ 4.3, 4.4), the numbers H, $\varkappa \ge 0$ and the functions $\eta(z)$, $\mu(z)$, $(0 \le z \le H)$, $f(\xi)$ $(-\infty < \xi < \infty)$ are known in the problem 4.1.

Using the complete energy expression, we can show (similarly as in § 1.1) that the solution $u(z,\xi)$ of the boundary-value problem 4.1 is unique. It exists under the conditions of theorems formulated below. Set

$$\frac{1}{\bar{v}^2(z)} = \frac{\eta(z)}{\mu(z)}, \qquad \bar{\sigma}(z) = \frac{\bar{v}(z)}{\mu(z)}, \qquad f(\xi) = \frac{\mathrm{d}\varphi}{\mathrm{d}\xi} \quad (\varphi(0) = 0),$$

where the function $\bar{v}(z)$ is positive, and change the variables

$$x = \int_0^z \frac{\mathrm{d}z}{\bar{v}(z)}, \qquad \sigma(x) = \bar{\sigma}(z).$$

4.2. SOLUTION OF THE DIRECT PROBLEM

The following theorem is used essentially when we solve the inverse problems in § 4.3, 4.4.

Theorem 4.2.1. *Suppose that the following conditions hold*

(A) the derivatives $\mathrm{d}^2\eta/\mathrm{d}z^2$ *and* $\mathrm{d}^2\mu/\mathrm{d}z^2$ *are continuous for* $0 \le z \le H$*;*

(B) *in the point* $z = H$, *we have* $\mathrm{d}\bar{\sigma}(z)/\mathrm{d}z = 0$;

(C) *the function* $\varphi(\xi)$ *and its derivatives up to the fourth order are continuous and absolutely integrable in the line* $-\infty < \xi < \infty$.

Then, for each $\varkappa > 0$, *the solution* $u(z,\xi)$ *of the direct problem 4.1 has the form*

$$u(z,\xi) = \int_{-\infty}^{\infty} U(z,k)e^{ik\xi}\,\mathrm{d}k, \qquad 0 \le z \le H, \quad -\infty < \xi < \infty, \tag{4.2.1}$$

$$U(z,k) = \bar{\sigma}^{1/2}(z)W(x,k), \qquad x = \int_0^z \frac{\mathrm{d}z}{\bar{v}(z)}, \qquad \sigma(x) = \bar{\sigma}(z), \tag{4.2.2}$$

$$W(x,k) = \frac{\gamma ik\tilde{\varphi}(k)\{\varphi'(x_H,k)\psi(x,k) - \psi'(x_H,k)\varphi(x,k)\}}{\varphi'(x_H,k) + ik\beta\psi'(x_H,k)}, \tag{4.2.3}$$

$$(0 \le x \le x_H, \qquad -\infty < k < \infty).$$

Here $x_H = x(H)$; *the functions* $\varphi(x,k)$ *and* $\psi(x,k)$ $(0 \le x \le x_H)$ *solve the equation*

$$-y'' + q(x)y = k^2 y, \qquad 0 \le x \le x_H,$$

where

$$q(x) = \sigma^{1/2}(x)\{\sigma^{-1/2}(x)\}'', \tag{4.2.4}$$

with the initial data

$$\varphi(0,k) = 1, \qquad \varphi'(0,k) = \alpha, \tag{4.2.5}$$

$$\psi(0,k) = 0, \qquad \psi'(0,k) = 1 \tag{4.2.6}$$

respectively;

$$\alpha = -\frac{1}{2}\frac{\sigma'(0)}{\sigma(0)}, \qquad \beta = \varkappa\bar{v}(0), \qquad \gamma = \sigma^{-1/2}(0)\bar{v}(0); \tag{4.2.7}$$

$\tilde{\varphi}(k)$ *is the Fourier transform of the function* $\varphi(\xi)$:

$$\varphi(\xi) = \int_{-\infty}^{\infty} \tilde{\varphi}(k)e^{ik\xi}\,\mathrm{d}k. \tag{4.2.8}$$

The integrals in formula (4.2.1) and in the equality

$$\frac{\partial u}{\partial z} = \int_{-\infty}^{\infty} \frac{\mathrm{d}U}{\mathrm{d}z}\, e^{ik\xi}\,\mathrm{d}k$$

are the Fourier integrals (i.e., their inversion formulas hold). Here and further in this chapter the prime denotes the differentiation with respect to x.

4.2.1. Reduction of the two-dimensional direct problem 4.1 in the strip $0 \le z \le H$, $-\infty < \xi < \infty$ to the one-dimensional problem in the finite segment $0 \le x \le x_H$ in the complex plane. We seek for the solution of the direct problem in form (4.2.1) where the integral is taken along the real axis $\operatorname{Im} k = 0$ of the complex plane $k = k_1 + ik_2$.

Suppose that the conditions (A), (B) of Theorem 4.2.1 hold. Substituting formally functions (4.2.1) and (4.2.8) into the problem 4.1 and changing the variables (4.2.2), we obtain the following problem for the function $W(x, k)$:

$$\frac{\mathrm{d}^2 W}{\mathrm{d}x^2} + \{k^2 - q(x)\}W = 0, \qquad 0 \le x \le x_H, \tag{4.2.9}$$

$$\frac{\mathrm{d}W}{\mathrm{d}x}(0, k) - (\alpha + ik\beta)W(0, k) = \gamma ik\tilde{\varphi}(k) \quad \text{for} \quad x = 0, \tag{4.2.10}$$

$$\frac{\mathrm{d}W}{\mathrm{d}x}(x_H, k) = 0 \quad \text{for} \quad x = x_H. \tag{4.2.11}$$

The function $W(x, k)$ must be analytic in the half-plane $\operatorname{Im} k < 0$, continuous up to the real axis $\operatorname{Im} k = 0$, bounded, and decrease for $|k| \to \infty$ in the half-plane $\operatorname{Im} k \le 0$ sufficiently fast[i)] for each fixed $x \in [0, x_H]$ (condition (4.2.12)).

Equation (4.2.9) and conditions (4.2.10), (4.2.11) are given in the real axis $-\infty < k < \infty$ of the plane k. The function $q(x)$ and the numbers α, β, γ are real and satisfy formulas (4.2.4), (4.2.7).

System (4.2.9)–(4.2.11) does not correspond to the self-adjoint Sturm—Liouville operator, since the coefficient $(\alpha + ik\beta)$ in condition (4.2.10) is complex and variable (depends on k).

4.2.2. The functions $\varphi(x, k)$ and $\psi(x, k)$. The function $q(x)$ is continuous in the segment $[0, x_H]$ by the condition (A), and equation (4.2.9) has the solutions $\varphi(x, k)$ and $\psi(x, k)$ respectively for initial data (4.2.5), (4.2.6) with the properties enumerated below (see Marchenko, 1952; Faddeev, 1959; Levitan and Sargsjan, 1970). For each fixed $x \in [0, x_H]$, $\varphi(x, k)$ and $\psi(x, k)$ are entire analytic functions with respect to k (and on $\lambda = k^2$, also); they are continuous with respect to (x, k) for $x \in [0, x_H]$ and any k; they form the fundamental system of solutions of equation (4.2.9); their Wronskian is as follows:

$$\psi'(x, k)\varphi(x, k) - \psi(x, k)\varphi'(x, k) \equiv 1. \tag{4.2.13}$$

[i)]The estimate for the function $W(x, k)$, sufficient for convergence of integral (4.2.1) and its derivatives, depends on the conditions imposed on the function $\varphi(\xi)$ and will be obtained below

Their asymptotic properties were studied by Marchenko (1952), Faddeev (1959), Levitan and Sargsjan (1970) and are as follows:

set $k = k_1 + ik_2$. Then there exists $k_0 > 0$, such that, for $|k| > k_0$, the following estimates hold:

$$\varphi(x,k) = \cos kx + O\Big(\frac{e^{|k_2|x}}{|k|}\Big), \qquad \psi(x,k) = \frac{\sin kx}{k} + O\Big(\frac{e^{|k_2|x}}{|k|}\Big).$$

4.2.3. Solution of problem (4.2.9)–(4.2.12). We shall seek the solution of problem (4.2.9)–(4.2.12) in the form

$$W(x,k) = c_1(k)\varphi(x,k) + c_2(k)\psi(x,k).$$

Defining $c_1(k)$, $c_2(k)$ from boundary conditions (4.2.10) and (4.2.11), we obtain formula (4.2.3) for $W(x,k)$.

We show now that the function $W(x,k)$ of form (4.2.3) satisfies condition (4.2.12). In order that the function $W(x,k)$ be analytic for $\operatorname{Im} k < 0$ and continuous for $\operatorname{Im} k \le 0$, $k \ne 0$, it is necessary and sufficient that the function

$$A(k) = \varphi'(x_H,k) + ik\beta\psi'(x_H,k)$$

does not vanish for $\operatorname{Im} k \le 0$, $k \ne 0$.

The scheme of the proof of this statement is as follows.

Parallel with the function $A(k)$, consider in a domain D_0 of the half-plane $\operatorname{Im} k \le 0$ the set of the functions

$$A_{k_0}(k) = \varphi'(x_H,k) + ik_0\beta\psi'(x_H,k)$$

dependent on the parameter $k_0 \in D_0$. Then, we establish the dependence of the zeros of the function $A_{k_0}(k)$ on the position of k_0 (more exactly, on the sign of the imaginary part of $ik_0\beta$). To this end, we connect the zeros of $A_{k_0}(k)$ with eigenvalues of a certain not self-adjoint operator (of the auxiliary bounded-value problem). This dependence is such that each function $A_{k_0}(k)$ from the set can not coincide with the correspondent number k_0 in the parameter $(ik_0\beta)$ (the zeros $A_{k_0}(k)$ are slipping from the number k_0, when k_0 runs in the half-plane $\operatorname{Im} k < 0$). As for the function $A(k)$, its zeros and the number k in the coefficient $(ik\beta)$ must coincide. Therefore, the function $A(k)$ has no zeros in the domain D_0[ii].

[ii] Analogous approach is used for another problem in Chapter 1. However, its establishment in our case is not evident and must be justified.

4.2.4. The auxiliary not self-conjugate boundary-value problem. Consider the following boundary-value problem in the segment $[0, x_H]$:

$$l(y) \equiv -y'' + q(x)y = k^2 y, \qquad 0 \le x \le x_H, \tag{4.2.14}$$

$$y'(0) - \theta_0 y(0) = 0, \tag{4.2.15}$$

$$y'(x_H) = 0. \tag{4.2.16}$$

Here θ_0 is a fixed complex number (independent of k), $q(x)$ is a complex continuous function.

Theorem 4.2.2. *Suppose $y(x)$ is a nontrivial solution of boundary-value problem (4.2.14)–(4.2.16) for a certain $\lambda = k^2$, where $y(x)$, $y'(x)$ belong to $L_2[0, x_H]$. Set*

$$\lambda = \lambda_1 + i\lambda_2, \quad \theta_0 = \theta_{01} + i\theta_{02}, \quad q(x) = q_1(x) + iq_2(x),$$

where $\lambda_1, \lambda_2, \theta_{01}, \theta_{02}, q_1(x), q_2(x)$ are real. Then, the identity

$$\lambda_2 \int_0^{x_H} |y|^2 \, dx = \theta_{02}|y(0)|^2 + \int_0^{x_H} q_2|y|^2 \, dx$$

holds.

This theorem yields the next theorem.

Theorem 4.2.3. *Suppose that one of the two following conditions holds*

$$\operatorname{Im}\theta_0 < 0, \qquad \operatorname{Im} q(x) \le 0 \qquad (0 \le x \le x_H),$$

or

$$\operatorname{Im}\theta_0 > 0, \qquad \operatorname{Im} q(x) \ge 0 \qquad (0 \le x \le x_H);$$

then, the eigenvalues of boundary-value problem (4.2.14)–(4.2.16), *i. e. the squares of zeros of the function*

$$A_{\theta_0}(k) = \varphi'(x_H, k) + (\theta - \alpha)\psi'(x_H, k)$$

may be situated only in the half-plane $\operatorname{Im} k^2 < 0$ *or* $\operatorname{Im} k^2 > 0$ *respectively.*

Theorems 4.2.2 and 4.2.3 are similar to those theorems for a certain not self-adjoint differential operator in the half-axis (see, Naymark, 1954, Subsection 3.7)[iii].

[iii]One of these theorems (Theorem 1.2.2) is formulated in Subsection 1.2.4.

Remark 4.2.1. Theorem 4.2.2 holds also if, instead of condition (4.2.16), we take the condition $y(x_H) = 0$, which corresponds for our problem to condition $u(H, \xi) = 0$ (fixed end) instead of condition (4.1.3). Therefore, Theorem 4.2.3 holds also for the function

$$B_{\theta_0}(k) = \varphi(x_H, k) + (\theta_0 - \alpha)\psi(x_H, k).$$

For the not self-adjoint operator, correspondent to boundary-value problem (4.2.14)–(4.2.16), the complex number θ_0 is fixed, and in system (4.2.9)–(4.2.11) the coefficient $(\alpha + ik\beta)$ depends on k. Therefore, we shall prove the following auxiliary statement.

4.2.5. Connection between the disposition of zeros of certain sets of functions $A_{\theta_0}(k)$ and the lack of zeros of the function $A(k)$.

We denote $\theta(k) = \alpha + ik\beta$ and consider a certain domain D_0 in the half-plane $\operatorname{Im} k < 0$. Parallel with the domain we consider the following domain D_c. Changing the parameter k_0 in the domain D_0, we consider the set of functions $\{A_{\theta_0}(k)\}_{D_0}$ dependent on the parameter

$$\theta_0 = \theta(k_0) = \alpha + ik_0\beta, \qquad k_0 \in D_0,$$

(where $\varphi(x, k)$ and $\psi(x, k)$ for these functions we shall take as for the function $A(k)$). Choosing an arbitrary point k_0 from the domain D_0, we fix a function $A_{\theta_0}(k)$ from this set and consider the disposition of zeros of this function which belong to the half-plane $\operatorname{Im} k < 0$. To this end, we apply Theorem 4.2.3. Suppose that all these zeros lie in the domain $D^c_{\theta_0}$ of the half-plane $\operatorname{Im} k < 0$.

Analogous constructions we shall do for each point $k_0 \in D_0$. The union of all these domains $D^c_{\theta_0}$, where $\theta_0 = \theta(k_0)$, $k_0 \in D_0$ we denote by D_c (D_c belongs evidently to the half-plane $\operatorname{Im} k < 0$).

The following theorem holds.

Theorem 4.2.4. *If the domains D_0 and D_c are not intersecting, then the function $A(k)$ has no zeros in the domain D_0.*

Using Theorems 4.2.3 and 4.2.4 we can show that the function $A(k)$ has no zeros in the quadrants $\operatorname{Im} k < 0$, $\operatorname{Re} k > 0$ and $\operatorname{Im} k < 0$, $\operatorname{Re} k < 0$, which we take in turn as the domain D_0. It is essential that the number β is positive (see (4.2.7)). It occurs that the function $\theta(k) = \alpha + ik\beta$, $\beta > 0$, is such that the correspondent domains D_0 and D_c are not intersecting.

Denote the half-axis $k_1 = 0$; $k_2 < 0$ by D_0. Suppose $L_{\theta_0,H}$ is the operator generated by the differential expression

$$l(y) = -y'' + q(x)y, \qquad 0 \le x \le x_H,$$

where $q(x)$ is taken from equation (4.2.9), and by the boundary conditions (4.2.15), (4.2.16). Each self-adjoint operator $L_{\theta_0,H}$ from the set $\{L_{\theta_0,H}\}_{D_0}$, which depends on the parameter $\theta_0 = \theta(k_0)$, $k_0 \in D_0$, is positive definite. This can be shown analogously to Subsection 1.2.6 or to paper of Alekseev (1962). Hence it follows that $A(k)$ has no zeros in the imaginary half-axis $k_2 < 0$.

If $-\infty < k < \infty$, the functions $\varphi'(x_H, k)$ and $\psi'(x_H, k)$ are real. As identity (4.2.13) holds, they are not equal to zero simultaneously. Therefore, in the half-axes $k_1 > 0$ and $k_1 < 0$, there are no zeros of the function $A(k)$ $(\beta \neq 0)$.

Further, taking into account the condition (B) and the explicit formulas

$$\varphi(x,0) = \sigma^{1/2}(0)\sigma^{-1/2}(x),$$
$$\psi(x,0) = \sigma^{-1/2}(0)\sigma^{-1/2}(x)\int_0^x \sigma(x)\,\mathrm{d}x, \tag{4.2.17}$$

we obtain that the point $k = 0$ is ordinary zero of the function $A(k)$; so, $W(x,k)$ is continuous in the point $k = 0$.

The asymptotic properties of the functions $\varphi(x,k)$ and $\psi(x,k)$ and certain considerations yield that the function $W(x,k)$is bounded for $|k| \to \infty$ in the domain $x \in [0, x_H]$, $\operatorname{Im} k \le 0$, and

$$W(x,-k) = \bar{W}(x,k), \qquad |W^{(n)}(x,k)| < Mk^n|\varphi(k)|,$$
$$n = 0,1,2, \qquad 0 \le x \le x_H, \qquad -\infty < k < \infty.$$

Applying these relations and the known theorem on connection of smoothness of $\varphi(\xi)$ and the velocity of decreasing its Fourier transform $\tilde{\varphi}(k)$, we obtain Theorem 4.2.1 as the result.

4.3. THE INVERSE PROBLEM WITH THE DATA IN THE FREE END OF THE STRING

Inverse problem 4.1. Suppose that for $0 \le \xi < \infty$ we are given the function $f(\xi)$ and the values $u(H,\xi)$ of the solution $u(z,\xi)$ of the direct problem 4.1 with unknown length H of the segment $[O,H]$, unknown number $\varkappa$ and unknown (for $0 \le z \le H$) functions $\eta(z)$ and $\mu(z)$. We know one of the three numbers $\sigma(0)$, $\sigma(H)$, $\varkappa\mu(0)$ (the other two numbers are unknown).

We suppose also that $\varkappa > 0$.

We need to find: in the general case, the numbers x_H, $\varkappa$, and the function $\sigma(x)$ for $0 \le x \le x_H$; if $\mu(z) \equiv \mu(0)$, then we must find the numbers H, $\varkappa$ and the function $\eta(z)$ for $0 \le z \le H$; if the function $\bar{v}^2(z) = \mu(z)/\eta(z)$ is known for $0 \le z \le H$, then we need to find the number $\varkappa$ and the functions $\eta(z)$ and $\mu(z)$ for $0 \le z \le H$.

The method of solution and the uniqueness theorem for the inverse problem. Let $q(x)$ be a certain real function continuous in a finite segment $x \in [0, l]$ and

$$L(y) = -y'' + q(x)y$$

is the correspondent differential expression. Denote by $S\{q(x), l, \varphi, \theta\}$ $(-\infty < \varphi, \theta < \infty)$ the set of eigenvalues (spectrum) of the boundary-value problem

$$L(y) = \lambda y,$$
$$\cos\varphi y(0) + \sin\varphi y'(0) = 0, \qquad \cos\theta y(l) + \sin\theta y'(l) = 0. \tag{4.3.1}$$

As is known, the number l is determined if a certain spectrum $S\{q(x), l, \varphi, \theta\} = \{\lambda_n\}_1^\infty$ is set, since (see Krein, 1951)

$$l^2 = \pi^2 \lim_{n\to\infty} \frac{n^2}{\lambda_n}. \tag{4.3.2}$$

Following the results of Borg (1945), Chudov (1949), Marchenko (1952), the spectra $S\{q(x), l, \varphi_1, \theta\}$ and $S\{q(x), l, \varphi_2, \theta\}$, $\varphi_1 \ne \varphi_2$ correspondent to two boundary-value problems with the same boundary condition in the point $x = l$ and different boundary conditions in the point $x = 0$, define both the function $q(x)$ and the numbers $\operatorname{ctg}\varphi_1$, $\operatorname{ctg}\varphi_2$, and $\operatorname{ctg}\theta$.

Boundary-value problem (4.3.1) is the classic (regular) Sturm—Liouville problem correspondent to a self-adjoint operator. It occurs that the data of the inverse problem 4.1 correspondent to *one* not self-adjoint system (4.2.9)–(4.2.11) allow us to find the spectra of *two* boundary self-adjoint problems of type (4.3.1) with the same condition in the point $x = x_H$ and different conditions in the point $x = 0$. We mean the problems

$$-y'' + q(x)y = \lambda y, \qquad 0 \le x \le x_H,$$
$$\cos\varphi y(0) + \sin\varphi y'(0) = 0, \qquad y'(x_H) = 0 \qquad (\operatorname{ctg}\varphi = -\alpha); \tag{I}$$

$$-y'' + q(x)y = \lambda y, \qquad 0 \le x \le x_H,$$
$$y(0) = 0, \qquad y'(x_H) = 0, \tag{II}$$

where the function $q(x)$ and the number α are taken from system (4.2.9)–(4.2.11).

Really, inverting the integral

$$u(H,\xi) = -\sigma^{1/2}(H) \int_{-\infty}^{\infty} \frac{\gamma ik\tilde{\varphi}(k)}{\varphi'(x_H,k) + ik\beta\psi'(x_H,k)} e^{ik\xi}\, dk$$

(use Theorem 4.2.1 and identity (4.2.13)), we find firstly the function

$$cA(k) = c\varphi'(x_H,k) + cik\beta\psi'(x_H,k) \qquad (c = \sigma^{-1/2}(H)\gamma^{-1}). \tag{4.3.3}$$

Then, separating its real and imaginary parts (the numbers c, β and the functions $\varphi(x,k), \psi(x,k)$ are real for $k \in (-\infty, \infty)$ are real and $\beta \ne 0$); namely, the functions

$$c\varphi'(x_H,k) \qquad \text{and} \qquad c\beta\psi'(x_H,k) \tag{4.3.4}$$

defined in the whole real axis $-\infty < k < \infty$. Therefore, we define all real zeros of the functions $\varphi'(x_H,k)$ and $\psi'(x_H,k)$. Their remaining zeros, as it follows from the properties of self-adjoint Sturm—Liouville operators (Gasymov and Levitan, 1964; Levitan and Sargsjan, 1970), may be purely imaginary only. However, the operators correspondent to the problems (I) and (II) are positive definite. This can be shown similarly as in Subsection 1.2.6. Therefore, the functions $\varphi'(x_H,k)$ and $\psi'(x_H,k)$ has no zeros which are purely imaginary, and the real zeros obtained are all the zeros of these functions.

Note that the squares $\lambda = k^2$ of all zeros of the function $\varphi'(x_H,k)$ and $\psi'(x_H,k)$ form the spectra $S\{q(x), x_H, \varphi, \pi/2\}$ and $S\{q(x), x_H, 0, \pi/2\}$ of boundary-value problems (I) and (II) respectively. Now, using these spectra, we find uniquely the desired function $q(x)$, $0 \le x \le x_H$ and the number α (the number x_H we define by formula (4.3.2)). For this purpose we apply the methods of Marchenko (1952), Levitan (1962) (see also Gasymov and Levitan, 1964; Krein, 1951).

Taking into account the function $q(x)$ and the number α we find the function $\sigma(x)$ for $0 \le x \le x_H$ using the formulas (4.2.17), for example. If $\sigma(0)$ or $\sigma(H)$ is set then $\sigma(x)$ is obtained from formulas (4.2.17) immediately. If the number $\varkappa\mu(0)$ is given, then the value $\sigma(0)$ we find by the formula

$\sigma(0) = \beta/(\varkappa\mu(0))$ defining, at first, the number β from expressions (4.3.4). Further, we find the number $\varkappa$ by the formula

$$\varkappa = \beta c \sigma^{1/2}(x_H)\sigma^{-1/2}(0),$$

where the number c we define from expression (4.3.3).

We can find the number $\mu(0)$ also from the equalities $\mu(0) = \bar{v}(0)/\sigma(0)$, $\bar{v}(0) = \varphi(H, 0)/c$, and the number $\varkappa$ from the equality $\varkappa = \beta/\bar{v}(0)$.

If $\mu(z) \equiv \mu(0)$, then $\mu(0)\sigma(x) = \bar{v}(x)$; and, using the relations

$$z = \int_0^x \bar{v}(x)\,\mathrm{d}x, \qquad \bar{v}(z) = \bar{v}(x), \qquad H = z(x_H)$$

we obtain the number H and the functions $\bar{v}(z)$ and $\eta(z) = 1/\bar{v}^2(z)$ for $0 \le z \le H$. So, the inverse problem 4.1 is solved.

From the way of solving this problem follows

Theorem 4.3.1. *Suppose the conditions (A), (B), (C) of Theorem 4.2.1 hold. Then the inverse problem 4.1 has a unique solution for each $\varkappa > 0$.*

4.4. THE INVERSE PROBLEM WITH THE DATA SET IN THE BOUNDARY $z = 0$

Inverse problem 4.2. Suppose the values $u(0, \xi)$ and $\partial u/\partial z|_{z=0}$ of the solution $u(z, \xi)$ of the direct problem 4.1 are set for $0 \le \xi < \infty$. Let the numbers H, $\varkappa$ and the functions $\eta(z)$, $\mu(z)$ ($0 \le z \le H$) and $f(\xi)$ ($-\infty < \xi < \infty$) be unknown. One of the three numbers $\mu(0)$, $\sigma(0)$ and $\sigma(H)$ is known, the other two are unknown. Let $\varkappa > 0$.

We need to find: in the general case — the number x_H and the function $\sigma(x)$ for $0 \le x \le x_H$; if $\mu(z) \equiv \mu(0)$ ($\mu(0)$ is given), then we need to find the number H and the function $\eta(z)$ for $0 \le z \le H$; if the function $\bar{v}^2(z) = \mu(z)/\eta(z)$ is known for $0 \le z \le H$, then we must find the functions $\eta(z)$ and $\mu(z)$ for $0 \le z \le H$.

From formulas (4.2.1)–(4.2.3), for $z = 0$, we find

$$U(0, k) = -\frac{\sigma^{1/2}(0)\gamma i k \tilde{\varphi}(k)\psi'(x_H, k)}{\varphi'(x_H, k) + ik\beta\psi'(x_H, k)},$$

$$\frac{\mathrm{d}U}{\mathrm{d}z}(0, k) = \frac{\sigma^{1/2}(0)\gamma i k \tilde{\varphi}(k)\varphi'(x_H, k)}{\bar{v}(0)\{\varphi'(x_H, k) + ik\beta\psi'(x_H, k)\}},$$

$$-\infty < k < \infty.$$

Hence all zeros of the functions $\varphi'(x_H, k)$ and $\psi'(x_H, k)$are defined uniquely for $-\infty < k < \infty$; therefore, the spectra of the boundary-value problems (I) and (II)[iv] are defined also. Further, we solve this inverse problem similarly as the inverse problem 4.1. The following theorem holds.

Theorem 4.4.1. *Suppose, all the conditions of Theorem 4.2.1 hold. Then the inverse problem 4.2 has a unique solution for each* $\varkappa > 0$.

4.5. INVERSE PROBLEMS FOR THE STRING WITH THE FIXED END $z = H$

The problem obtained from the direct problem 4.1 changing condition (4.1.3) by the condition

$$u(H, \xi) = 0, \qquad -\infty < \xi < \infty$$

we shall call the direct problem 4.2.

Theorem 4.5.1. *Suppose the conditions (A) and (C) of Theorem 4.2.1 hold. Then, for each* $\varkappa > 0$, *the solution of the direct problem 4.2 is represented in the form*

$$u(z, \xi) = \int_{-\infty}^{\infty} U(z, k) e^{ik\xi}\, \mathrm{d}k, \tag{4.5.1}$$

$$U(z, k) = \sigma^{1/2}(z) \frac{\gamma ik\tilde{\varphi}(k)\{\varphi(x_H, k)\psi(x, k) - \psi(x_H, k)\varphi(x, k)\}}{\varphi(x_H, k) + ik\beta\psi(x_H, k)}. \tag{4.5.2}$$

The variable x, the numbers x_H, β, γ and the functions $\tilde{\varphi}(k)$, $\varphi(x, k)$, $\psi(x, k)$, $(0 \le x \le x_H)$ has the same sense as in Theorem 4.2.1.

The integral in equality (4.5.1) (and for the function $\partial u/\partial z$) is the Fourier integral (i. e. the inversion formulas hold).

The proof is analogous to those described in § 4.2 (see Remark 4.2.1).

Inverse problem 4.3. Suppose the conditions (A) and (C) of Theorem 4.2.1 hold and the function $f(\xi)$ and the values $\partial u/\partial z|_{z=H}$ are set for $0 \le \xi < \infty$. Here $u(z, \xi)$ is the solution of the direct problem 4.2 with unknown numbers H, $\varkappa$ and unknown functions $\eta(z)$ and $\mu(z)$ (for $0 \le z \le H$).

[iv] the spectra are obtained rather simply if we introduce the additional condition: the set of all real zeros of the function $\tilde{\varphi}(k)$ does not intersect with each of sets of zeros of the functions $\varphi'(x_H, k)$ and $\psi'(x_H, k)$. Then, since the numbers of two different spectra are strictly alternating (see Gasymov and Levitan, 1964), we have $S\{q(x), x_H, \varphi, \pi/2\} = U - U \cap D$, $S\{q(x), x_H, 0, \pi/2\} = D - U \cap D$, where U and D are the sets of squares of all real zeros of the functions $U(0, k)$ and $\partial U/\partial z(0, k)$ respectively.

Let, also, one of three numbers $\varkappa\mu(0)$, $\sigma(0)$, $\sigma(H)$ be set (two other numbers are unknown) and $\varkappa > 0$. We need to find: in the general case, the number x_H and the function $\sigma(x)$ for $0 \le x \le x_H$; if $\mu(z) \equiv \mu(0)$, then we seek the number H and the function $\eta(z)$ for $0 \le z \le H$; if the function $\bar{v}^2(z) = \mu(z)/\eta(z)$ is known for $0 \le z \le H$, then we need to find the functions $\eta(z)$ and $\mu(z)$ for $0 \le z \le H$.

Taking into account formulas (4.5.2), we have

$$\frac{\mathrm{d}U}{\mathrm{d}z}(H,k) = \frac{\sigma^{1/2}(H)}{\bar{v}(0)} \frac{\gamma ik\tilde{\varphi}(k)}{\varphi(x_H,k) + ik\beta\psi(x_H,k)}, \qquad -\infty < k < \infty.$$

Hence we find all (real) zeros of the functions $\varphi(x,k)$ and $\psi(x,k)$ and the spectra $S\{q(x), x_H, \varphi, 0\}$ and $S\{q(x), x_H, 0, 0\}$ of the boundary-value problems obtained from the problems (I) and (II), § 4.3 by the change of the condition $y'(x_H) = 0$ by $y(x_H) = 0$. Therefore, solution and the proof of the uniqueness theorem of the inverse problem 4.3 is just the same as those described in § 4.3.

Inverse problem 4.4. We suppose that the conditions (A) and (C) of Theorem 4.2.1 hold, and the values $u(0,\xi)$ and $\partial u/\partial z|_{z=0}$ of the solution $u(z,\xi)$ of the direct problem 4.2 are set for $0 \le \xi < \infty$. The numbers H, $\varkappa$ and the functions $\eta(z)$, $\mu(z)$ ($0 \le z \le H$), $f(\xi)$ ($-\infty < \xi < \infty$) are unknown. We know one of three numbers $\mu(0)$, $\sigma(0)$, $\sigma(H)$ (two other are unknown). Suppose, also, that $\varkappa > 0$. We need to find: in the general case, the number x_H and the function $\sigma(x)$ for $0 \le x \le x_H$; if $\mu(z) \equiv \mu(0)$ (the number $\mu(0)$ is given), then we must find the number H and the function $\eta(z)$ for $0 \le z \le H$; if the function $\bar{v}^2(z) = \mu(z)/\eta(z)$ is known for $0 \le z \le H$, then we have to find the functions $\eta(z)$ and $\mu(z)$ for $0 \le z \le H$.

From the data of this inverse problem we find the spectra $S\{q(x), x_H, \varphi, 0\}$ and $S\{q(x), x_H, 0, 0\}$. Then, following § 4.3 and § 4.4 we prove the uniqueness of its solution.

Chapter 5.

Inverse problems for the elliptic equation in the strip

We consider the problem of determination of variable coefficients (and the number H) in the elliptic equation

$$\frac{\partial^2 u}{\partial z^2} + \eta(z)\frac{\partial^2 u}{\partial \xi^2} + \tau(z)\frac{\partial u}{\partial z} = 0 \qquad (0 < m < \eta(z) < M < \infty), \tag{5.0.1}$$

considered in the strip $0 \le z \le H$, $-\infty < \xi < \infty$. The boundary-value problem (direct) for this equation is as follows: in the boundary $z = 0$ of the strip, we have the inhomogeneous condition with the directional derivative; in the boundary $z = H$, we have the condition $\partial u/\partial z|_{z=H} = 0$ (or, $u(H,\xi) = 0$). The data on the solution of the direct problem are usually as follows:

1) either the values of $u(z,\xi)$ set in the boundary $z = H$ (or the values $\partial u/\partial z|_{z=H}$).

2) or the values $u(z,\xi)$ and $\partial u/\partial z(z,\xi)$ in the boundary $z = 0$.

In both cases the inverse problem is solved similarly. Firstly, we construct the solution of the direct problem by the method of separation of variables (using Theorems 4.2.3, 4.2.4). The one-dimensional operator in this case is not self-adjoint and is not Sturm—Liouville. Nevertheless, using the obtained representation of the direct problem the inverse problems for equation (5.0.1) are reduced to the known problem of determination of the Sturm—Liouville equation given the two spectra. This problem already deals with the self-adjoint operator, which allows us to solve it and to prove the uniqueness theorem for these inverse problems.

The analogous approach is used for solution of the inverse problems for the hyperbolic equation in Chapter 4. However, in this case, unlike the method of Chapter 4, we have to apply the method of analytic extension for solution of the inverse problem.

5.1. SETTING OF THE DIRECT PROBLEM

Boundary-value (direct) problems 5.1. We consider

$$\frac{\partial^2 u}{\partial z^2} + \eta(z)\frac{\partial^2 u}{\partial \xi^2} + \tau(z)\frac{\partial u}{\partial z} = 0 \qquad (0 < m < \eta(z) < M < \infty), \tag{5.1.1}$$

in the domain $0 < z < H$, $-\infty < \xi < \infty$. The boundary condition

$$\left(\frac{\partial u}{\partial z} - \varkappa\frac{\partial u}{\partial \xi}\right)\Big|_{z=0} = f(\xi), \quad (-\infty < \varkappa < \infty) \tag{5.1.2}$$

is given for $z = 0$, $-\infty < \xi < \infty$ and the condition

$$\frac{\partial u}{\partial z}\Big|_{z=H} = 0 \tag{5.1.3}$$

is given in for $z = H$, $-\infty < \xi < \infty$.

Suppose also that the following conditions hold.

The function $u(z, \xi)$ and its derivatives u_z, u_ξ are continuous in the strip $0 \le z \le H$, $-\infty < \xi < \infty$. The derivatives of the second order are continuous in the segment $0 < z < H$, $-\infty < \xi < \infty$.

The function $u(z, \xi) \to 0$ when $|\xi| \to \infty$ uniformly relatively z in the interval $[0, H]$.

The coefficients of equation (5.1.1) are bounded for $0 < z < H$.

We need to find the function $u(z, \xi)$ defined in the strip $0 \le z \le H$, $-\infty < \xi < \infty$ which satisfies all the conditions mentioned above.

We shall call this problem by the boundary-value (direct) problem 5.1.

Taking into account the extremum principle of Hopf and Zaremba—Giraud (Bitsadze, 1966) (see, also, Hopf, 1927, 1952; Zaremba, 1927; Giraud, 1932; Oleinik, 1952; Miranda, 1957), we can show that the only one solution of the boundary-value problem 5.1 may exist. The method of the proof is similar to considerations of § 2.1. The solution of the boundary-value problem 5.1 exists under the conditions of Theorem 5.2.1 which will be formulated below.

We introduce the positive functions $\mu(z)$, $\overset{+}{v}(z)$, and $\sigma(z)$ by the formulas

$$\tau(z) = \frac{\mathrm{d}}{\mathrm{d}z}\{\ln\mu(z)\} \quad (\mu(0) = 1), \qquad \eta(z) = \frac{1}{\overset{+}{v}{}^2(z)}, \qquad \overset{+}{\sigma}(z) = \frac{\overset{+}{v}(z)}{\mu(z)}.$$

Set

$$f(\xi) = \frac{\mathrm{d}\varphi}{\mathrm{d}\xi} \qquad (\varphi(0) = 0)$$

and change the variables

$$x = \int_0^z \frac{\mathrm{d}z}{\overset{+}{v}(z)}, \qquad \sigma(x) = \overset{+}{\sigma}(z).$$

5.2. SOLUTION OF THE DIRECT PROBLEM

The solution of the inverse problems of § 5.3, 5.4 uses essentially the explicit form of solution of the direct problem 5.1, which gives the following theorem.

Theorem 5.2.1. *Suppose that*

(A) the derivatives $\mathrm{d}^2\eta/\mathrm{d}z^2$ and $\mathrm{d}\tau/\mathrm{d}z$ are continuous for $0 \le z \le H$;

(B) $\mathrm{d}\overset{+}{\sigma}(z)/\mathrm{d}z = 0$ in the point $z = H$.

(C) the function $\varphi(\xi)$ and its derivatives up to fourth order are continuous and absolutely integrable in the line $-\infty < \xi < \infty$.

Then the solution $u(z,\xi)$ of the direct problem 5.1 has the form

$$u(z,\xi) = \int_{-\infty}^{\infty} U(z,k)e^{ik\xi}\,\mathrm{d}k, \quad 0 \le z \le H, \quad -\infty < \xi < \infty, \tag{5.2.1}$$

$$U(z,k) = \overset{+}{\sigma}{}^{1/2}(z)W(x,k), \quad x = \int_0^z \frac{\mathrm{d}z}{\overset{+}{v}(z)}, \quad \sigma(x) = \overset{+}{\sigma}(z), \tag{5.2.2}$$

$$W(x,k) = \frac{\gamma ik\tilde{\varphi}(k)\{\varphi'(x_H,ik)\psi(x,ik) - \psi'(x_H,ik)\varphi(x,ik)\}}{\varphi'(x_H,ik) + ik\beta\psi'(x_H,ik)},$$
$$0 \le x \le x_H, \qquad -\infty < k < \infty. \tag{5.2.3}$$

Here $x_H = x(H)$; the functions $\varphi(x,s)$ and $\psi(x,s)$ $(0 \le x \le x_H)$ solve the equation

$$-y'' + q(x)y = s^2 y, \qquad 0 \le x \le x_H,$$

where

$$q(x) = \sigma^{1/2}(x)\{\sigma^{-1/2}(x)\}'' \tag{5.2.4}$$

with the initial data respectively

$$\varphi(0,s) = 1, \qquad \varphi'(0,s) = \alpha; \tag{5.2.5}$$

$$\psi(0,s) = 0, \qquad \psi'(0,s) = 1; \tag{5.2.6}$$

$$\alpha = -\frac{1}{2}\frac{\sigma'(0)}{\sigma(0)}, \qquad \beta = \varkappa\sigma(0), \qquad \gamma = \sigma^{1/2}(0); \tag{5.2.7}$$

$\tilde{\varphi}(k)$ is the Fourier transform of the function $\varphi(\xi)$:

$$\varphi(\xi) = \int_{-\infty}^{\infty} \tilde{\varphi}(k)e^{ik\xi}\,\mathrm{d}k. \tag{5.2.8}$$

The integrals in formula (5.2.1) and in the equality

$$\frac{\partial u}{\partial z} = \int_{-\infty}^{\infty} \frac{\partial U}{\partial z}\,e^{ik\xi}\,\mathrm{d}k$$

are the Fourier integrals (i. e., the inversion formulas hold). Here and below in Chapter 5 the prime denotes the differentiation with respect to x.

The proof of Theorem 5.2.1 is analogous to approach of § 4.2 and is based on Theorems 4.2.3, 4.2.4.

Proof. The solution $u(z,\xi)$ of the direct problem 5.1 we seek in form (5.2.1), where the integral is taken along the real axis of the complex plane $k = k_1 + ik_2$. Suppose the conditions of Theorem 5.2.1 hold. Substituting the functions from (5.2.1), (5.2.8) into the conditions of the direct problem 5.1, changing the variables (5.2.2), and passing to the new variable s by the formulas

$$s = ik, \qquad W(x,s) = W(x,k), \tag{5.2.9}$$

we obtain the following problem for the function $W(x,s)$:

$$\frac{\mathrm{d}^2 W}{\mathrm{d}x^2} + \{s^2 - q(x)\}W = 0, \qquad 0 \le x \le x_H, \tag{5.2.10}$$

$$\frac{\mathrm{d}W}{\mathrm{d}x}(0,s) - (\alpha + s\beta)W(0,s) = \gamma ik\tilde{\varphi}(k) \quad \text{for} \quad x = 0, \tag{5.2.11}$$

$$\frac{\mathrm{d}W}{\mathrm{d}x}(x_H, s) = 0 \quad \text{for} \quad x = x_H. \tag{5.2.12}$$

The function $W(x,s)$ must be continuous, bounded in the imaginary axis $\operatorname{Re} s = 0$, $-\infty < \operatorname{Im} s < \infty$ and decrease sufficiently fast[i)] for $|\operatorname{Im} s| \to \infty$, $\operatorname{Re} s = 0$ for each fixed $x \in [0, x_H]$ (condition (5.2.13)).

The real function $q(x)$ and the numbers α, β, γ are expressed by formulas (5.2.4), (5.2.7).

Unlike the problem for the hyperbolic equation from Chapter 4, equation (5.2.10) and conditions (5.2.11) and (5.2.12) are set not in the real axis, but on the imaginary axis $\operatorname{Re} s = 0$ of the complex plane. The solution $W(x,s)$ of problem (5.2.10)–(5.2.13) must be defined, therefore, for all purely imaginary values s and for $x \in [0, x_H]$. Similarly as for the problem from Chapter 4 for the string equation, system (5.2.10)–(5.2.12) does not correspond to self-adjoint regular Sturm—Liouville operator, since the coefficient $(\alpha + s\beta)$ in condition (5.2.11) is complex and depends on s.

By the condition (A), the function $q(x)$ is continuous for $0 \le x \le x_H$ and equation (5.2.10) has the solutions $\varphi(x,s)$ and $\psi(x,s)$ respectively, for initial data (5.2.5), (5.2.6), defined for all complex s and $x \in [0, x_H]$. These solutions are entire functions with respect to s for fixed $x \in [0, x_H]$ (their properties see in Subsection 4.2.2). Using these functions we defined the solution $W(x,s)$ of system (5.2.10)–(5.2.12) in the form

$$W(x,s) = \frac{\gamma s \tilde{\varphi}(k)\{\varphi'(x_H,s)\psi(x,s) - \psi'(x_H,s)\varphi(x,s)\}}{\varphi'(x_H,s) + s\beta\psi'(x_H,s)},$$
$$-\infty < \operatorname{Im} s < \infty, \quad \operatorname{Re} s = 0; \quad 0 \le x \le x_H. \tag{5.2.14}$$

In order that the function $W(x,s)$ of form (5.2.14) be continuous and bounded in the imaginary axis $\operatorname{Re} s = 0$ for $s \neq 0$, it is necessary and sufficient that the function

$$A(s) = \varphi'(x_H,s) + s\beta\psi'(x_H,s)$$

has no zeros in this axis for $s \neq 0$. The proof of the property that $A(s)$ has no zeros in the set

$$\operatorname{Re} s = 0, \qquad -\infty < \operatorname{Im} s < \infty; \qquad s \neq 0$$

is similar to those in Subsection 4.2.5.

[i)]The estimate for the function $W(x,s)$ sufficient for convergence of integral (5.2.1) and its derivatives depends on the conditions imposed on the function $\varphi(\xi)$ and is formulated below.

Denote the imaginary half-axis $s_2 > 0$ by D_0. Parallel with $A(s)$ ($s \in D_0$) we consider the set of the functions

$$A_{\theta_0}(s) = \varphi'(x_H, s) + (\theta_0 - \alpha)\psi'(x_H, s)$$

dependent on the parameter $\theta_0 = \alpha + s_0\beta$, where $s_0 \in D_0$. For each function $A_{\theta_0}(s)$ from this set $\{A_{\theta_0}(s)\}_{D_0}$ we have

$$\operatorname{Im}\theta_0 = s_{02}\beta > 0 \qquad (\text{if} \quad \beta > 0).$$

By Theorem 4.2.3, each function $A_{\theta_0}(s)$ may have zeros only for $\operatorname{Im} s^2 = 2s_1s_2 > 0$, which corresponds to the quadrant $s_2 > 0$, $s_1 > 0$ in the half-plane $s_2 > 0$ (the domain D_c in terms of Theorem 4.2.4). As $D_0 \cap D_c = 0$, then, by Theorem 4.2.4, the function $A(s)$ has no zeros in the set D_0. This holds also and for $\beta < 0$.

When $\beta = 0$, we have $A(s) = \varphi'(x_H, s)$; as it was noticed in § 4.3 this function has no zeros which are purely imaginary. Condition (B) and formulas (4.2.17) yield that the point $s = 0$ is the prime zero of the function $A(s)$; therefore, $W(x, s)$ is continuous in the point $s = 0$, $0 \le x \le x_H$.

Returning by formulas (5.2.9) to the complex plane k, using the relations

$$W(x, -k) = \bar{W}(x, k), \qquad |W^{(n)}(x, k)| < Mk^n|\tilde{\varphi}(k)|,$$
$$n = 0, 1, 2, \qquad -\infty < k < \infty, \qquad 0 \le x \le x_H$$

and the connection between the smoothness of $\varphi(\xi)$ and decreasing $\tilde{\varphi}(k)$, we obtain Theorem 5.2.1 as the result. □

5.3. THE INVERSE PROBLEM WITH THE DATA IN THE BOUNDARY $z = H$

Inverse problem 5.1. Let the function $f(\xi)$ and $u(H, \xi)$ be given for $-\infty < \xi < \infty$. Here $u(z, \xi)$ is the solution of the direct problem 5.1; the number H and the functions $\eta(z)$, $\tau(z)$ ($0 \le z \le H$) are unknown. One of the numbers $\sigma(0)$, $\sigma(H)$, $\varkappa$ is known (the two other numbers are unknown). Suppose also that $\varkappa \ne 0$.

We need to find: in the general case, the number x_H and the function $\sigma(x)$ for $0 \le x \le x_H$; if $\tau(z) \equiv 0$, then we must find the number H and the function $\eta(z)$ for $0 \le z \le H$; if the function $\eta(z)$ is known for $0 \le z \le H$, then we find the function $\tau(z)$ for $0 \le z \le H$. If $\varkappa$ is not given, we find $\varkappa$ also.

Parallel with differential system (5.2.10)–(5.2.12) we consider the following two boundary-value regular Sturm—Liouville problems with the same boundary conditions in the point $x = x_H$ and different boundary conditions in the point $x = 0$, correspondent to the self-adjoint operators

$$\begin{gathered} -y'' + q(x)y = \lambda y, \qquad 0 \leq x \leq x_H, \\ \cos\varphi y(0) + \sin\varphi y'(0) = 0, \qquad y'(x_H) = 0 \qquad (\operatorname{ctg}\varphi = -\alpha), \end{gathered} \tag{I}$$

$$\begin{gathered} -y'' + q(x)y = \lambda y, \qquad 0 \leq x \leq x_H, \\ y(0) = 0, \qquad y'(x_H) = 0, \end{gathered} \tag{II}$$

where the function $q(x)$ and the number α are taken from system (5.2.10)–(5.2.12) (which does not correspond to a self-adjoint Sturm—Liouville operator).

We reduce the inverse problem 5.1 (similar to the inverse problems of Chapter 4 for the hyperbolic equation) to the problem of determination of the Sturm—Liouville operator given the two spectra. From the data of the inverse problem 5.1 we find the spectra $S\{q(x), x_H, \varphi, \pi/2\}$ and $S\{q(x), x_H, 0, \pi/2\}$ of the boundary-value problems (I) and (II) respectively (we use the notations § 4.3). However, this inverse problem is differed essentially from the inverse problems of Chapter 4 by the way of determination of these spectra.

Indeed, we invert the integral

$$u(H,\xi) = -\sigma^{1/2}(H) \int_{-\infty}^{\infty} \frac{\gamma ik\tilde{\varphi}(k)}{\varphi'(x_H, ik) + ik\beta\psi'(x_H, ik)}\, e^{ik\xi}\, dk.$$

Here we apply Theorem 5.2.1 and the identity

$$\varphi(x,s)\psi'(x,s) - \varphi'(x,s)\psi(x,s) \equiv 1$$

and define, at first, the function

$$c\varphi'(x_H, ik) + ik\beta\psi'(x_H, ik) \quad (c = \sigma^{-1/2}(H)\gamma^{-1})$$

for $-\infty < k < \infty$. Then we separate its real and imaginary parts (the numbers c, β and the functions $\varphi(x,s), \psi(x,s)$ are real in the imaginary axis $\operatorname{Re} s = 0$ and $\beta \neq 0$). We find the two functions

$$c\varphi'(x_H, s) \qquad \text{and} \qquad c\beta\psi'(x_H, s) \tag{5.3.1}$$

in the imaginary axis $\operatorname{Re} s = 0$, $-\infty < \operatorname{Im} s < \infty$ of the complex plane (not in the real axis, as in Chapter 4).

However, the functions $\varphi'(x_H, s)$ and $\psi'(x_H, s)$ are entire functions with respect to s. Using this property, we extend analytically the functions (5.3.1) (the numbers c, β are unknown), which are defined in a certain (may be extremely small) interval of the imaginary axis $\operatorname{Re} s = 0$ on the whole real axis $-\infty < s < \infty$. To this end, we apply the Taylor expansion of the functions (5.3.1).

In this case, we define all the real zeros of the functions $\varphi'(x_H, s)$ and $\psi'(x_H, s)$; therefore [ii], all zeros of these functions.

Note that the squares $\lambda = s^2$ of zeros of these functions $\varphi'(x_H, s)$ and $\psi'(x_H, s)$ (disposed as the increasing sequence $\lambda_1, \lambda_2, \dots, \lambda_n, \dots$) form the spectra

$$S\{q(x), x_H, \varphi, \pi/2\} \qquad \text{and} \qquad S\{q(x), x_H, 0, \pi/2\}$$

of the boundary-value problems (I) and (II) respectively. Following the results of Borg (1945), Chudov (1949) and Marchenko (1952); these spectra uniquely yield the function $q(x)$ for $0 \le x \le x_H$ and the number α. Note that the number x_H we find with the aid of one of these spectra using the known (Gasymov and Levitan, 1964) formula

$$x_H = \pi^2 \lim_{n \to \infty} \frac{n^2}{\lambda_n}.$$

Applying the methods (see, Krein, 1951; Gasymov and Levitan, 1964; Marchenko, 1952) of determination of the regular Sturm—Liouville operator given two spectra, we find the function $q(x)$ for $0 \le x \le x_H$ and the number α. Then, analogous to Chapter 4, we find the function $\sigma(x)$ for $0 \le x \le x_H$ and the number $\varkappa$ (if it is unknown).

From the method of solution of this inverse problem follows the theorem.

Theorem 5.3.1. *Let the conditions (A), (B), (C) of Theorem 5.2.1 hold. Then the inverse problem 5.2 has a unique solution for each* $\varkappa \neq 0$.

5.4. THE INVERSE PROBLEM WITH THE DATA IN THE BOUNDARY $z = 0$

Inverse problem 5.2. Let, for $-\infty < \xi < \infty$, be given the values $u(0, \xi)$ and $\partial u / \partial z|_{z=0}$. Here $u(z, \xi)$ is the solution of the direct problem 5.1 with

[ii] As was noticed in § 4.3, these functions have no other zeros.

unknown numbers $H, \varkappa$ and unknown functions $\eta(z), \tau(z)$ $(0 \le z \le H)$, $f(\xi)$ $(-\infty < \xi < \infty)$.

We need to find: in the general case, the number x_H and the function $\sigma(x)$ for $0 \le x \le x_H$; if $\tau(z) \equiv 0$, then the number H and the function $\eta(z)$ for $0 \le z \le H$; if the function $\eta(z)$ is known for $0 \le z \le H$, then we need to find the function $\tau(z)$ for $0 \le z \le H$.

Applying Theorem 5.2.1 we find the functions

$$U(0,k) = -\sigma^{1/2}(0)\,\frac{\gamma ik\tilde{\varphi}(k)\psi'(x_H, ik)}{\varphi'(x_H, ik) + ik\beta\psi'(x_H, ik)},$$

$$\frac{\mathrm{d}U}{\mathrm{d}z}(0,k) = \sigma^{-1/2}(0)\,\frac{\gamma ik\tilde{\varphi}(k)\varphi'(x_H, ik)}{\varphi'(x_H, ik) + ik\beta\psi'(x_H, ik)}$$

for $-\infty < k < \infty$. These formulas and the method of analytical extension yield the spectra

$$S\{q(x), x_H, \varphi, \pi/2\} \qquad \text{and} \qquad S\{q(x), x_H, 0, \pi/2\}$$

of the boundary-value problems (I) and (II) uniquely. Then, the solution of the inverse problem follows from § 5.3. So, the following statement holds.

Theorem 5.4.1. *Let the conditions of Theorem 5.2.1 hold. Then, the inverse problem 5.2 has a unique solution for each real $\varkappa$.*

5.5. PROBLEMS WITH THE CONDITION $u(H,\xi) = 0$

The problem, which is obtained from the direct problem 5.1 changing condition (5.1.3) by the condition

$$u(H,\xi) = 0, \qquad -\infty < \xi < \infty$$

we call by the direct problem 5.2. From the Hopf and Zaremba–Girand principles follow that this problem may have the only one solution $u(z,\xi)$.

The role of Theorem 5.2.1 here plays the following statement.

Theorem 5.5.1. *Suppose that the condition (A) and (C) of Theorem 5.2.1 hold. Then the solution $u(z,\xi)$ of the direct problem 5.2 has the representation*

$$u(z,\xi) = \int_{-\infty}^{\infty} U(z,k)e^{ik\xi}\,\mathrm{d}k, \qquad 0 \le z \le H, \quad -\infty < \xi < \infty, \qquad (5.5.1)$$

$$U(z,k) = \overset{+}{\sigma}{}^{1/2}(z)\,\frac{\gamma ik\tilde{\varphi}(k)\{\varphi(x_H,ik)\psi(x,ik) - \psi(x_H,ik)\varphi(x,ik)\}}{\varphi(x_H,ik) + ik\beta\psi(x_H,ik)}.$$

The variable x, the numbers x_H, β, γ and the functions $\tilde{\varphi}(k)$, $\varphi(x,s)$, $\psi(x,s)$ $(0 \le x \le x_H)$ have the same meaning that in Theorem 5.2.1. The integral in equality (5.5.1) *(and for the function $(\partial u)/(\partial z)$ also) is the Fourier integral.*

The proof of this theorem is analogous to those used in § 5.2 and is based on Remark 4.2.1 and Theorem 4.2.4.

Inverse problem 5.3. Let the conditions (A) and (C) of Theorem 5.2.1 hold. Suppose that the functions $f(\xi)$, $\partial u/\partial z|_{z=H}$, and $u(z,\xi)$ are known for $-\infty < \xi < \infty$, where $u(z,\xi)$ is the solution of the direct problem 5.2 with unknown number H and the functions $\eta(z)$, $\tau(z)$ $(0 \le z \le H)$. Suppose, also, that one of the numbers $\sigma(0)$, $\sigma(H)$, $\varkappa$ is known (and two others are unknown). Suppose also that $\varkappa \ne 0$.

We need to find: in the general case, the number x_H and the function $\sigma(x)$ for $0 \le x \le x_H$; if $\tau(z) \equiv 0$, then we need to find the number H and the function $\eta(z)$ for $0 \le z \le H$; if the function $\eta(z)$ is known for $0 \le z \le H$, then we must find the function $\eta(z)$ for $0 \le z \le H$. Also, we find the number $\varkappa$ if it is unknown.

Using Theorem 5.5.1 we find the functions

$$c\varphi(x_H,s) \qquad \text{and} \qquad c\beta\psi(x_H,s) \quad (c = \sigma^{1/2}(H)\gamma^{-1}) \tag{5.5.2}$$

in the imaginary axis of the complex plane S. Extending analytically entire functions (5.5.2) in the real axis $-\infty < s < \infty$, we find all the zeros of the functions $\varphi(x_H,s)$ and $\psi(x_H,s)$ and the spectra

$$S\{q(x), x_H, \varphi, 0\} \qquad \text{and} \qquad S\{q(x), x_H, 0, 0\}$$

of the boundary-value problems. These problems are obtained from the problems (I) and (II) by replacing the condition $y'(x_H) = 0$ by the condition $y(x_H) = 0$. Further solution and the proof of the uniqueness theorem for the inverse problem 5.3 follows similar to the methods of § 5.3.

Inverse problem 5.4. Suppose that the conditions (A) and (C) of Theorem 5.2.1 hold. Suppose also that the values $u(0,\xi)$ and $\partial u/\partial z|_{z=0}$ are set for $-\infty < \xi < \infty$, where $u(z,\xi)$ is the solution of the direct problem 5.2 with unknown numbers H, $\varkappa$ and unknown functions $\eta(z)$, $\tau(z)$ $(0 \le z \le H)$, $f(\xi)$ $(-\infty < \xi < \infty)$.

We need to find: in the general case, the number x_H and the function $\sigma(x)$ for $0 \le x \le x_H$; if $\tau(z) \equiv 0$, then we seek for the number H and the function $\eta(z)$ for $0 \le z \le H$; if the function $\eta(z)$ is known for $0 \le z \le H$, then we must find the function $\tau(z)$ for $0 \le z \le H$.

Taking into account the inverse problem data, applying Theorem 5.5.1 and the method of analytical extension, we find the spectra

$$S\{q(x), x_H, \varphi, 0\} \qquad \text{and} \qquad S\{q(x), x_H, 0, 0\}$$

and, following § 5.3, 5.4, we establish the uniqueness of its solution.

Chapter 6.

Inverse problems of scattering the plane waves from inhomogeneous layers with a free or fixed boundary

This chapter is devoted to problems where characteristics of inhomogeneous elastic layer with free (or fixed) boundary are determined.

Similarly the problem of Chapter 3 for inhomogeneous half-space (transition layer), the problem of scattering the plane waves of SH-type from the inhomogeneous elastic layer with free (or fixed) boundary, whose properties are described by an arbitrary function dependent on the normal coordinate, is reduced to a certain two-dimensional problem in the strip with the directional derivative in one of the boundary conditions. Depending on the character of inhomogeneous layer and the angle of incidence of the plane wave the differential equation of this problem may have hyperbolic, elliptic and mixed types. In this chapter we consider the inverse problems for first two cases. The case of the mixed type equation will be investigated in Chapter 7.

Firstly, for both cases: hyperbolic and elliptic, we formulate the problems of determining the properties of inhomogeneous layer given either the vibration in a certain point of free (fixed) boundary and the form of incident wave or the forms of incident and reflected waves set (in both problems) for a fixed angle of incidence. Uniqueness theorems and the way of solution of these inverse problems are obtained directly carrying over the results of Chapter 4 and 5.

In the general case, we may determine uniquely only a certain intermediate characteristic of the layer, the function $\sigma(x,\theta_0)$; the density ρ and the speed of the wave propagation v as the function of depth z is remained unknown. Therefore, in § 6.3, we suggest the way of recovery of the depth of the inhomogeneous layer and the functions $v(z)$ and $\rho(z)$ using the data of each of inverse problems formulated above, given for a certain set of values of the incidence angle θ_0 of the plane wave.

However, if certain additional conditions are imposed on the form of incident wave, we succeed (in both hyperbolic and elliptic cases) in formulation of the inverse problem (in § 6.4), where the depth of inhomogeneous layer, the density $\rho(z)$, the speed $v(z)$ and the form of incident wave — all this is determined uniquely. The data, in this case, is the set of signal records of a fixed point of free (or fixed) boundary dependent on the angle of incidence θ_0. Similar to the inverse problems with the data for a single value of θ_0, this inverse problem is reduced (in both cases) to the problem of determining the regular Sturm—Liouville operator given two spectra. Unlike the inverse problems formulated before, the spectra here have the other meaning.

In the inverse problems with the data for one value θ_0 (§ 6.2), we find the spectra of the regular Sturm—Liouville problems which play the role of frequency for a fixed value θ_0 of the parameter. In the problems of § 6.4, conversely, we find the spectra of the Sturm—Liouville problems given the parameter θ_0 for certain fixed value of frequency. The determination is based on using the analyticity of the functions with respect to θ_0 by means of which the solution of the direct problem is expressed. Besides, unlike to inverse problems with the data for one angle θ_0, here, the solution of the problem of determining the Sturm—Liouville operator given the two spectra yields immediately the values of H and the functions $v(z)$ and $\rho(z)$.

Note that in all the problems of this chapter the characteristics $v(z)$ and $\rho(z)$ may be not monotone.

6.1. THE DIRECT PROBLEM

Suppose that in the domain $0 < z < H$, $-\infty < \xi < \infty$, the hyperbolic equation

$$\frac{\partial}{\partial z}\left\{\mu(z)\,\frac{\partial u}{\partial z}\right\} = \eta(z)\,\frac{\partial^2 u}{\partial \xi^2}, \tag{6.1.1}$$

is considered, where

$$\eta(z) = \frac{\mu(z)}{\bar{v}^2(z,\theta_0)}, \qquad \frac{1}{\bar{v}^2(z,\theta_0)} = \frac{1}{v^2(z)} - \frac{\sin^2\theta_0}{v_0^2}. \tag{6.1.1'}$$

For $z = 0$ the boundary condition with the directional derivative is given; for $z = H$, we consider the condition of free boundary

$$\left(\frac{\partial u}{\partial z} - \varkappa\frac{\partial u}{\partial \xi}\right)\Big|_{z=0} = \frac{\partial \varphi}{\partial \xi}, \tag{6.1.2}$$

$$\frac{\partial u}{\partial z}\Big|_{z=H} = 0, \qquad -\infty < \xi < \infty, \tag{6.1.3}$$

where

$$\varkappa = \frac{\delta}{\mu(0)}, \qquad \delta = \cos\theta_0 \frac{\mu_0}{v_0}, \qquad \varphi(\xi) = -2\varkappa\varphi_0(\xi,\theta_0); \tag{6.1.2'}$$

besides, the initial condition is given

$$u(z,\xi,\theta_0)\Big|_{\xi<0} \equiv 0, \qquad z \geq 0. \tag{6.1.4}$$

The functions $v(z), \mu(z)$ and the constants v_0, μ_0 are positive. Each of the conditions

$$v(0) = v_0, \quad v(0) \neq v_0; \qquad \mu(0) = \mu_0, \quad \mu(0) \neq \mu_0;$$

is admissible; the function $\varphi_0(\xi,\theta_0)$ has the property

$$\varphi_0(\xi,\theta_0)\,|_{\xi<0} = 0.$$

We need to find the function $u(z,\xi,\theta_0)$ satisfying conditions (6.1.1)–(6.1.4).

The problem of scattering the plane waves SH from the inhomogeneous elastic layer with the boundary free from stress in the case

$$v(z) < v_0/\sin\theta_0, \qquad 0 \leq z \leq H \tag{H}$$

is reduced to problem (6.1.1)–(6.1.4). This case we shall naturally call hyperbolic. We can establish this fact similarly as in § 3.1 (see Megrabov, 1973a). Here $v(z)$ and $\mu(z)$ are the speed of wave propagation and the modulus shift in the inhomogeneous layer $0 \leq z \leq H$ respectively; v_0 and μ_0 are the correspondent constants in the homogeneous layer $z < 0$, whence the plane wave of the form $\varphi_0(\xi,\theta_0)$ is falling ($-\infty < \xi < \infty$). The angle of incidence θ_0

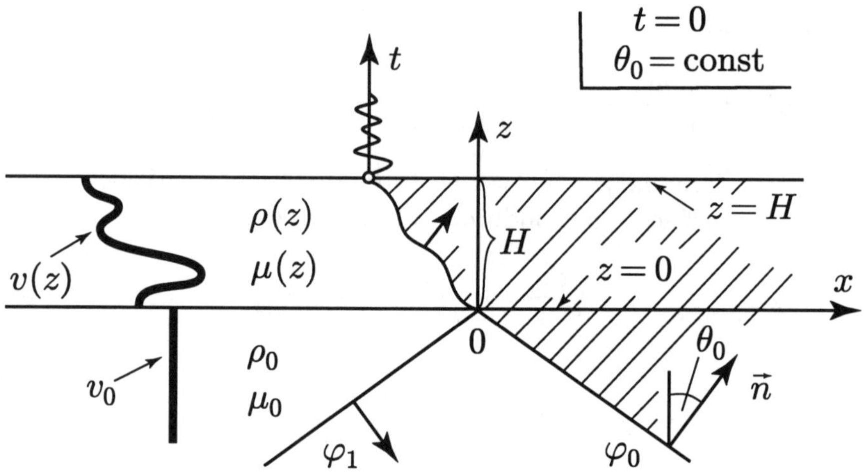

Figure 6.1. To the problem of scattering the plane waves from inhomogeneous layers with the free or fixed boundary in the coordinates x, z, t

$(0 \le \theta_0 < \pi/2)$ is given; $u(z, \xi, \theta_0) = u(x, z, t, \theta_0)$, where $u(x, z, t, \theta_0)$ is the summary shift field in the layer; $\xi = t - x \sin\theta_0/v_0$. The choice of the coordinate system x, z, t (t is physical time) is clear from Figure 6.1.

Problem (6.1.1)–(6.1.4), evidently, is a case of the boundary-value problem 4.1 for a finite string. If the function $u(z, \xi) = u(z, \xi, \theta_0)$ is sufficiently smooth (these conditions are formulated in § 1, Chapter 3), problem (6.1.1)–(6.1.4) has a unique solution $u(z, \xi)$. This problem we call by the direct problem 6.1.

All the results of § 4.2 and Theorem4.2.1 are carrying over the direct problem 6.1.

Consider now the case

$$v(z) > v_0/\sin\theta_0, \qquad 0 \le z \le H, \tag{E}$$

when the differential equation in the reduced two-dimensional problem has the elliptic type everywhere in the strip $0 \le z \le H$, $-\infty < \xi < \infty$.

Introduce the positive function $\overset{+}{v}(z, \theta_0)$ by the formula

$$\frac{1}{\overset{+}{v}{}^2(z, \theta_0)} = \frac{\sin^2\theta_0}{v_0^2} - \frac{1}{v^2(z)}, \qquad 0 \le z \le H.$$

In the elliptic case (E) we obtain the following direct problem.

In the domain $0 < z < H$, $-\infty < \xi < \infty$, we consider the equation

$$\frac{\partial^2 u}{\partial z^2} + \eta(z)\frac{\partial^2 u}{\partial \xi^2} + \tau(z)\frac{\partial u}{\partial z} = 0 \qquad (0 < m < \eta(z) < M < \infty), \qquad (6.1.5)$$

where

$$\tau(z) = \frac{\mathrm{d}}{\mathrm{d}z}\{\ln \mu(z)\}, \qquad \eta(z) = \frac{1}{\overset{+}{v}{}^2(z, \theta_0)}.$$

For $z = 0$ and $z = H$, we have the boundary conditions

$$\left(\frac{\partial u}{\partial z} - \varkappa\frac{\partial u}{\partial \xi}\right)\bigg|_{z=0} = f(\xi), \qquad (6.1.6)$$

$$\frac{\partial u}{\partial z}\bigg|_{z=H} = 0, \qquad -\infty < \xi < \infty, \qquad (6.1.7)$$

where

$$\varkappa = \frac{\delta}{\mu(0)}, \qquad \delta = \cos\theta_0 \frac{\mu_0}{v_0}, \qquad f(\xi) = -2\varkappa\frac{\mathrm{d}\varphi_0}{\mathrm{d}\xi}.$$

Suppose also that

$$u(z, \xi) \to 0 \quad \text{for} \quad |\xi| \to \infty \qquad (6.1.8)$$

uniformly relatively $z \in [0, H]$. To conditions (6.1.5)–(6.1.8) we add the smoothness conditions imposed on the function $u(z, \xi) = u(z, \xi, \theta_0)$ formulated in the boundary-value problem 5.1. We need to find the function $u(z, \xi)$ satisfying all the conditions formulated above.

This problem we shall call the direct problem 6.2. The problem 6.2 is a particular case of the direct problem 5.1; therefore, it may have the only one solution $u(z, \xi)$. All the results of § 5.2 and Theorem 5.2.1 are carried over the direct problem 6.2.

In both cases the form $\varphi_1(\xi, \theta_0)$ of the plane wave reflected from the inhomogeneous layer, being unknown in the direct problem, is connected with the solution $u(z, \xi, \theta_0)$ by the boundary condition

$$u(0, \xi, \theta_0) = \varphi_0(\xi, \theta_0) + \varphi_1(\xi, \theta_0). \qquad (6.1.9)$$

6.2. DETERMINATION OF PROPERTIES OF THE INHOMOGENEOUS LAYER GIVEN THE DATA FOR A SINGLE ANGLE OF INCIDENCE

Introduce in the hyperbolic case the function $\sigma(x,\theta_0)$ obtained from the function

$$\bar{\sigma}(z,\theta_0) = \mu(z)/\bar{v}(z,\theta_0)$$

by the change of variables

$$x = \int_0^z \frac{dz}{\bar{v}(z,\theta_0)}, \qquad \sigma(x,\theta_0) = \bar{\sigma}(z,\theta_0) \equiv \sigma(x(z,\theta_0),\theta_0).$$

In the elliptic case we introduce the function $\overset{+}{\sigma}(x,\theta_0)$ obtained from the function

$$\overset{+}{\sigma}(z,\theta_0) = \mu(z)/\overset{+}{v}(z,\theta_0)$$

by the change of variables

$$x = \int_0^z \frac{dz}{\overset{+}{v}(z,\theta_0)}, \quad \overset{+}{\sigma}(x,\theta_0) = \overset{+}{\sigma}(z,\theta_0) \equiv \overset{+}{\sigma}(x(z,\theta_0),\theta_0).$$

Inverse problem 6.1 (with the data on the free boundary).

Suppose that for $-\infty < \xi < \infty$, and for a certain $\theta_0 \in [0,\pi/2)$, in both cases: hyperbolic and elliptic, the function $\varphi_0(\xi,\theta_0)$ (the form of incident wave) is set and the values $u(H,\xi)$ of the solution $u(z,\xi)$ are set. Here $u(z,\xi)$ is the solution of the direct problem 6.1 and 6.2 respectively with unknown (in both cases) numbers H, v_0, μ_0 and the functions $v(z)$, $\mu(z)$ $(0 \le z \le H)$. Also, in both cases, one of the numbers $\cos\theta_0\mu_0/v_0$, $\sigma(0,\theta_0)$ or $\overset{+}{\sigma}(0,\theta_0)$, $\sigma(H,\theta_0)$, or $\overset{+}{\sigma}(H,\theta_0)$ is given (all other numbers are unknown).

In both cases we need to find in the general case, for θ_0 given, the number $x_H = x(H)$ and the function $\sigma(x,\theta_0)$ or $\overset{+}{\sigma}(x,\theta_0)$ for $0 \le x \le x_H$ respectively. If $\mu(z) \equiv \mu(0)$ and v_0 given, we need to find the number H (the layer depth) and the function $v(z)$ for $0 \le z \le H$. If the number v_0 and the function $v(z)$ for $0 \le z \le H$ are given, then we need to find the function $\mu(z)$ for $0 \le z \le H$. In all the cases we need to find the number $\cos\theta_0\mu_0/v_0$ if it is unknown.

In the hyperbolic case this problem is equivalent to the inverse problem 4.1, and in the elliptic case it is equivalent to the inverse problem 5.1. Therefore, all the results of § 4.3 and § 5.3 are transformed to this problem and the following theorem holds.

Theorem 6.2.1. *Let the following conditions hold*

(A) the derivatives $\mathrm{d}^2 v/\mathrm{d}z^2$ *and* $\mathrm{d}^2\mu/\mathrm{d}z^2$ *are continuous for* $0 \le z \le H$;

(B) in the point $z = H$, *we have either* $\mathrm{d}\bar{\sigma}/\mathrm{d}z = 0$ *or* $\mathrm{d}\overset{+}{\sigma}/\mathrm{d}z = 0$;

(C) for θ_0 *given, the function* $\varphi_0(\xi, \theta_0)$ *and its derivatives with respect to* ξ *up to the fourth order are continuous and absolutely integrable in the line* $-\infty < \xi < \infty$.

Then in both hyperbolic and elliptic cases the inverse problem 6.1 has a unique solution.

Inverse problem 6.2 (determination of the properties of the layer given the forms of incident and reflected waves).

Let, for $-\infty < \xi < \infty$ and $\theta_0 \in [0, \pi/2)$ fixed, be given the functions $\varphi_0(\xi, \theta_0)$ and $\varphi_1(\xi, \theta_0)$ in both hyperbolic and elliptic cases. Here $\varphi_1(\xi, \theta_0)$ is connected by condition (6.1.9) with the solution $u(z, \xi, \theta_0)$ of the direct problem 6.1 and 6.2 respectively with unknown numbers H, v_0, μ_0 and the functions $v(z), \mu(z)$ $(0 \le z \le H)$ (in both cases). Also, one of four numbers is given: $\cos\theta_0\mu_0/v_0$, $\mu(0)$, $\sigma(0, \theta_0)$ or $\overset{+}{\sigma}(0, \theta_0)$, $\sigma(H, \theta_0)$ or $\overset{+}{\sigma}(H, \theta_0)$ (the other three numbers are unknown).

In both hyperbolic and elliptic cases we need to find: in the general case, for θ_0 given, we need to find the number $x_H = x(H)$ and the function $\sigma(x, \theta_0)$ or $\overset{+}{\sigma}(x, \theta_0)$ for $0 \le x \le x_H$ respectively. If $\mu(z) \equiv \mu(0)$ and the numbers $\mu(0)$, v_0 are known then we must find the number H and the function $v(z)$ for $0 \le z \le H$. If the number v_0 and the function $v(z)$ $(0 \le z \le H)$ are known, then we find $\mu(z)$ for $0 \le z \le H$.

Analogous to § 4, Chapter 4 and § 4, Chapter 5 this problem can be reduced to the problem of determination of the Sturm—Liouville operator given the two spectra. So, the following theorem holds.

Theorem 6.2.2. *Let the conditions of Theorem 6.2.1 hold. Then in both hyperbolic and elliptic cases the inverse problem 6.2 has a unique solution.*

6.3. DETERMINATION OF THE DEPTH OF INHOMOGENEOUS LAYER, THE DENSITY $\rho(z)$ AND THE SPEED $v(z)$ IN THIS LAYER IF THE FORM OF INCIDENT WAVE $\varphi_0(\xi, \theta_0)$ IS KNOWN

Inverse problem 6.3. Denote by $\Theta_0(\Theta_0^e)$ the infinite monotone increasing (or decreasing) sequence of real numbers $\theta_0^{(1)}, \theta_0^{(2)}, \ldots, \theta_0^{(n)}, \ldots$ such that

1) each θ_0 from the sequence Θ_0 belongs to the interval $[0, \pi/2)$ and satisfies the condition (H) ((E));
2) the sequence $\Theta_0(\Theta_0^e)$ converges to a certain limit point $\bar{\theta}_0 \neq 0$.

Let the conditions (A), (B) of Theorem 6.2.1 hold and the function $\varphi_0(\xi, \theta_0)$ satisfies the condition (C) of this theorem for each $\theta_0 \in \Theta_0(\Theta_0^e)$. Let for $-\infty < \xi < \infty$, for each $\theta_0 \in \Theta_0(\Theta_0^e)$ be given:

1) either the function $\varphi_0(\xi, \theta_0)$ and the values $u(H, \xi, \theta_0)$ of the solution $u(z, \xi, \theta_0)$ of the direct problem 6.1 or 6.2 respectively;

2) or the functions $\varphi_0(\xi, \theta_0)$ and $\varphi_1(\xi, \theta_0)$, where $\varphi_1(\xi, \theta_0)$ is connected with the solution $u(z, \xi, \theta_0)$ of the direct problem 6.1 and 6.2 by condition (6.1.9).

In both cases the number H and the functions $v(z)$ and $\mu(z)$ in the direct problems 6.1 or 6.2 are unknown. The numbers v_0 and μ_0 are given. Thus, the data of one of the inverse problems of § 2 for each $\theta_0 \in \Theta_0$ or $\theta_0 \in \Theta_0^e$ are given.

We need to find the number H and the functions $v(z)$ and $\mu(z)$ for $0 \leq z \leq H$.

Theorem 6.3.1. *Inverse problem 6.3 has a unique solution.*

Proof. The proof of this theorem and the way of solution of the inverse problem 6.3 are based on using formulas (3.3.3) and (3.5.1). Evidently, these formulas are independent of the domain shape where the differential equation is considered and on the boundary conditions. These formulas express only the connection between the coefficients we seek and the function $\sigma(x, \theta_0)$ or $\overset{+}{\sigma}(x, \theta_0)$. Therefore, at first, for each fixed $\theta_0 \in \Theta_0(\Theta_0^e)$, we solve the correspondent inverse problem § 6.2 and determine (for example, using the methods of Chapter 4, 5) the function $\sigma(x, \theta_0)$ or $\overset{+}{\sigma}(x, \theta_0)$. Then, using formulas (3.3.3) or (3.5.1), we recover the number H and the functions $v(z)$, $\mu(z)$ and $\rho(z) = \mu(z)/v^2(z)$. □

Today we have no numerical algorithm of solution of correspondent inverse problem § 6.2 with the data given for one value θ_0. If we have such an algorithm, we use formula (3.3.3) or (3.5.1) and derive the depth of inhomogeneous layer, the density $\rho(z)$ and the speed $v(z)$ numerically.

6.4. DETERMINATION OF THE DEPTH OF INHOMOGENEOUS LAYER, THE DENSITY $\rho(z)$, THE SPEED $v(z)$ IN THE LAYER AND THE FORM OF INCIDENT WAVE

The following theorem will be used below.

Theorem 6.4.1. *Let the conditions (A) and (C) of Theorem 6.2.1 hold and the condition*

(B′) in the point $z = H$ we have $\mathrm{d}v/\mathrm{d}z = 0$, $\mathrm{d}\mu/\mathrm{d}z = 0$

hold. Then, for each θ_0 satisfying the condition (H), *the solution of the direct problem 6.1 is represented in the form*

$$u(z,\xi,\theta_0) = \int_{-\infty}^{\infty} U(z,k,\theta_0)e^{ik\xi}\,\mathrm{d}k, \quad 0 \le z \le H, \quad -\infty < \xi < \infty, \tag{6.4.1}$$

$$\begin{aligned} U(z,k,\theta_0) = {} & 2\,\frac{\cos\theta_0}{\sigma_0}\,\mu^{-1/2}(0)\mu^{-1/2}(z)ik\tilde{\varphi}_0(k,\theta_0) \\ & \times \{\Phi(z,k,\theta_0)\Psi'(H,k,\theta_0) - \Psi(z,k,\theta_0)\Phi'(H,k,\theta_0)\} \\ & \Big/ \Big[\Phi'(H,k,\theta_0) + ik\,\frac{\cos\theta_0}{\sigma_0}\,\frac{1}{\mu(0)}\,\Psi'(H,k,\theta_0)\Big], \end{aligned}$$

$$0 \le z \le H, \qquad -\infty < k < \infty, \tag{6.4.2}$$

where $\sigma_0 = v_0/\mu_0$ and $\Phi(z,k,\theta_0)$ and $\Psi(z,k,\theta_0)$ are solutions of the equation

$$\frac{\mathrm{d}^2U}{\mathrm{d}z^2} + \Big\{p(z) + \frac{k^2}{\bar{v}^2(z,\theta_0)}\Big\}U = 0. \tag{6.4.3}$$

The function $p(z)$ in this equation is defined as follows:

$$p(z) = -\frac{1}{4}a^2(z) - \frac{1}{2}a'(z) = -\frac{1}{2}\frac{\mu''}{\mu} + \frac{1}{4}\frac{\mu'^2}{\mu^2}, \qquad a(z) = \frac{\mu'}{\mu} \tag{6.4.4}$$

respectively, with the initial data

$$\Phi(0,k,\theta_0) = 1, \qquad \Phi'(0,k,\theta_0) = \tilde{\alpha} = \frac{1}{2}\,a(0) = \frac{1}{2}\frac{\mu'(0)}{\mu(0)}, \tag{6.4.5}$$

$$\Psi(0,k,\theta_0) = 0, \qquad \Psi'(0,k,\theta_0) = 1. \tag{6.4.6}$$

Here $\tilde{\varphi}_0(k,\theta_0)$ is the Fourier transform of the function $\varphi_0(\xi,\theta_0)$. The prime here and further denotes the differentiation with respect to z. The integrals in formulas (6.4.1) and in the equality

$$\frac{\partial u}{\partial z} = \int_{-\infty}^{\infty} \frac{\mathrm{d}U}{\mathrm{d}z}\, e^{ik\xi}\, \mathrm{d}k$$

are the Fourier integrals.

This statement follows immediately from Theorem 4.2.1 and formulas (6.1.1′), (6.1.2′).

Lemma 6.4.1. *The functions $\Phi(z,k,\theta_0)$ and $\Psi(z,k,\theta_0)$ for each pair of fixed values $z \in [0,H]$ and $k^2 \in (-\infty,\infty)$ are entire analytic functions with respect to θ_0 and to the complex variable*

$$s = -k^2 \sin^2\theta_0 / v_0^2. \qquad (6.4.7)$$

Really, taking into account formulas (6.1.1′), (6.4.4), we can write equation (6.4.3) in the form

$$-\frac{\mathrm{d}^2 U}{\mathrm{d}z^2} + q(k,z)U = sU, \qquad 0 \le z \le H, \qquad (6.4.8)$$

where the function

$$q(k,z) = -\{p(z) + k^2/v^2(z)\}$$

is continuous by the condition (A). Up to this moment, the parameter θ_0 in the direct problem 6.1 was real. Let now θ_0 be complex. Then, for k fixed, the variable s is the entire function with respect to θ_0, and equation (6.4.8) is the regular Sturm—Liouville equation. It is known (see Levitan and Sargsyan, 1970, p. 14) that the solution of this equation are entire functions with respect to s if conditions (6.4.5)–(6.4.6) hold. Hence the lemma follows.

Note that the solution $u(z,\xi,\theta_0)$ of the direct problem 6.1 is defined by formulas (6.4.1), (6.4.2) only for real $\theta_0 \in [0,\pi/2)$ satisfying the condition (H). However, the functions $\Phi(z,k,\theta_0)$ and $\Psi(z,k,\theta_0)$ are defined for all complex values θ_0 and s.

Inverse problem 6.4. Consider the monotone decreasing (or increasing) sequence Θ_0 of real numbers $\theta_0^{(1)}, \theta_0^{(2)}, \dots, \theta_0^{(n)}, \dots$ such that

1) each θ_0 from the sequence Θ_0 belongs to the interval $[0,\pi/2)$ and satisfies the condition (H);

2) the sequence Θ_0 converges to a certain limit point $\bar{\theta}_0$ (evidently, this condition follows from the first condition). In the case of Θ_0 monotonically decreasing, the limit point may be, for example, $\theta_0 = 0$ (the case of normal falling).

Let the form of falling wave is independent of θ_0: $\varphi_0(\xi, \theta_0) \equiv \varphi_0(\xi)$.

We suppose that the conditions of Theorem 6.4.1 hold, and the values $u(H, \xi, \theta_0)$ of the solution $u(z, \xi, \theta_0)$ of the direct problem 6.1 are set. In this case the number H is unknown and the functions $v(z)$, $\mu(z)$ $(0 \le z \le H)$ and $\varphi_0(\xi)$ $(-\infty < \xi < \infty)$ are unknown. All these conditions hold for each value θ_0 from the sequence Θ_0. Besides, we are given the numbers v_0, μ_0 and the two real values $k = k_1$ and $k = k_2$ not equal to zero such that the numbers $\tilde{\varphi}_0(k_1)$ and $\tilde{\varphi}_0(k_2)$ ($\tilde{\varphi}_0(k)$ is the Fourier transform of the function $\varphi_0(\xi)$) are unknown. The numbers $\tilde{\varphi}_0(k_1)$ and $\tilde{\varphi}_0(k_2)$ are either real or pure imaginary (not necessarily simultaneously) and not equal to zero. We need to find the number H and the functions $v(z)$ and $\mu(z)$ for $0 \le z \le H$ and $\varphi_0(\xi)$ for $-\infty \le \xi \le \infty$.

Solution of the inverse problem 6.4.

Parallel with the direct problem 6.1, for k_1 given, we consider the two regular Sturm—Liouville problems with the same conditions in the point $z = H$ and different conditions in the point $z = 0$:

$$\begin{gathered} -y'' + q(k_1, z)y = sy, \qquad 0 \le z \le H, \\ \cos\varphi y(0) + \sin\varphi y'(0) = 0, \qquad y'(H) = 0 \quad (\operatorname{ctg}\varphi = -\tilde{\alpha}); \end{gathered} \tag{I}$$

$$\begin{gathered} -y'' + q(k_1, z)y = sy, \qquad 0 \le z \le H, \\ y(0) = 0, \qquad y'(H) = 0, \end{gathered} \tag{II}$$

where the numbers $H, \tilde{\alpha}$ and the functions $v(z)$ and $\mu(z)$ (in terms of which the function $q(k_1, z)$ is expressed) are taken from the direct problem 6.1.

Given k_1, v_0 and the sequence Θ_0, taking into account (6.4.7), we construct the sequence $s_1, s_2, \ldots, s_n, \ldots$ $(s_n = s(k_1, \theta_0^{(n)}))$. Denote this sequence by S_{k_1}. Evidently, S_{k_1} is converged to a finite limit $\bar{s}$. Taking into account formulas (6.4.1), (6.4.2) and the identity

$$\Phi\Psi' - \Phi'\Psi \equiv 1,$$

we obtain

$$u(H, \xi, \theta_0) = \int_{-\infty}^{\infty} \frac{2\mu^{-1/2}(0)\cos\theta_0 \mu^{-1/2}(H) ik\tilde{\varphi}_0(k)/\sigma_0}{\Phi'(H, k, \theta_0) + ik\cos\theta_0 \Psi'(H, k, \theta_0)/(\sigma_0\mu(0))} e^{ik\xi}\, dk. \tag{6.4.9}$$

Inverting the integral in (6.4.9) for $k = k_1$ and for each θ_0 from the sequence Θ_0, we define the values of the function

$$A(k_1, \theta_0) = c\{\Phi'(H, k_1, \theta_0) + ik_1 \cos\theta_0 \Psi'(H, k_1, \theta_0)/(\sigma_0\mu(0))\},$$
$$c = \mu^{1/2}(0)\mu^{1/2}(H)/(\sigma_0\tilde{\varphi}_0(k_1)) \qquad (6.4.10)$$

for each $\theta_0 \in \Theta_0$. Separating the real and imaginary parts of the function $A(k_1, \theta_0)$ we define the values of the functions $\{c\Phi'(H, k_1, \theta_0)\}$ and $\{c\beta\Psi'(H, k_1, \theta_0)\}$ for each $\theta_0 \in \Theta_0$, where $\beta = 1/(\sigma_0\mu(0))$ (the functions Φ and Ψ are real for real k and θ_0).

We shall consider the functions Φ and Ψ as the functions dependent on the variables $s = -k_1^2 \sin^2\theta_0/v_0^2$:

$$\Phi(z, k_1, \theta_0) = \Phi(z, k_1, s), \qquad \Psi(z, k, \theta_0) = \Psi(z, k_1, s).$$

By Lemma 6.4.1, the functions $\{c\Phi'(H, k_1, s)\}$ and $\{c\beta\Psi'(H, k_1, s)\}$ are entire functions with respect to s for k_1 given. We had found their values for all s from the sequence S_{k_1} converging to the finite limit $\bar{s}$. Therefore, by the uniqueness theorem for analytic functions, the functions $\{c\Phi'(H, k_1, s)\}$ and $\{c\beta\Psi'(H, k_1, s)\}$ are determined uniquely in the whole complex s-plane. Thus, all real zeros of the functions $\Phi'(H, k_1, s)$ and $\Psi'(H, k_1, s)$ are determined uniquely.

Note that the sets of all real zeros of the functions $\Phi'(H, k_1, s)$ and $\Psi'(H, k_1, s)$ form the spectra of the boundary-value problems (I) and (II) respectively. This follows from the general theory (see, Gasymov and Levitan, 1964; Levitan and Sargsyan, 1970, Chapter 1) of classical Sturm—Liouville problem. Following the results of G. Borg, L. A. Chudov, and M. A. Marchenko (see, Gasymov and Levitan, 1964) the spectra we have found define the number $\tilde{\alpha}$ and the function $q(k_1, z)$, $0 \le z \le H$, uniquely (the number H we may find using one of the spectra taking into account formula (4.3.2)). For this purpose we apply the methods of Gasymov and Levitan (1964), Levitan (1962), Krein (1951) and Marchenko (1952) of determination of the regular Sturm—Liouville operator, the two spectra be given.

We have not described the way of determining the functions $\{c\Phi'(H, k_1, s)\}$ and $\{c\beta\Psi'(H, k_1, s)\}$ in the axis $-\infty < s < \infty$ if their values are known for $s \in S_{k_1}$. Let the functions $u(H, \xi, \theta_0)$ be given not only for $\theta_0 \in \Theta_0$, but for all values θ_0 from a certain (extremely small) segment $0 \le \theta_1 \le \theta_0 \le \theta_2 < \pi/2$ satisfying the condition (H). Then, solving the inverse problem 6.4, we can extend analytically the functions $\{c\Phi'(H, k_1, s)\}$ and $\{c\beta\Psi'(H, k_1, s)\}$ from the segment $[s(k_1, \theta_1), s(k_1, \theta_2)]$ on the whole line

$-\infty < s < \infty$ by means of the Taylor expansion in a certain point of this segment.

Doing the same actions for $k = k_2$ and solving the second time the problem of determining the Sturm—Liouville operator, the two spectra given, we find the function $q(k_2, z)$ for $0 \le z \le H$. Then we derive the function $v(z)$ and $p(z)$ by the formulas:

$$v(z) = \left\{ \frac{k_2^2 - k_1^2}{q(k_2, z) - q(k_1, z)} \right\}^{1/2}, \qquad p(z) = -q(k_1, z) + \frac{k_1^2}{v^2(z)}$$

for $0 \le z \le H$. Taking into account the recovered number $\tilde{\alpha} = a(0)$ and (6.4.4), we find the function $a(z)$. Up to this moment we does not use the number μ_0. If μ_0 is given, then we find the value $\mu(0) = 1/\beta\sigma_0$ (defining β from expression (6.4.10) beforehand). Then, from equation (6.4.4) we derive the function $\mu(z)$ for $0 \le z \le H$.

Further, taking into account the recovered values H, $v(z)$, and $\mu(z)$, we construct the function $A(k, \theta_0)$ for a certain $\theta_0 \in \Theta_0$ and for all $k \in (-\infty, \infty)$. Inverting the integral in (6.4.9) for this value θ_0, we find the function

$$\tilde{\varphi}_0(k) = U(H, k, \theta_0) A(k, \theta_0) / 2\cos\theta_0 \quad \text{for} \quad -\infty < k < \infty,$$

and then we find the function $\varphi_0(\xi)$ for $-\infty < \xi < \infty$.

The following statement follows immediately from this way of solution of the inverse problem 6.4.

Theorem 6.4.2. *Inverse problem 6.4 has a unique solution* $\{H, v(z), \mu(z), \varphi_0(\xi)\}$.

If in the inverse problem 6.4, the form of incident wave is given then

1) the restrictions imposed on $\varphi_0(\xi)$, connected with existence of the numbers k_1 and k_2 are eliminated;

2) the function φ_0 may be dependent on θ_0: $\varphi_0 = \varphi_0(\xi, \theta_0)$, where $\varphi_0(\xi, \theta_0)$ for each fixed $\theta_0 \in \Theta_0$ must satisfy the condition (C) of Theorem 6.2.1.

The function φ_0 may be considered as dependent on θ_0 also in the inverse problem 6.4 with unknown function φ_0 if we introduce the following additional condition for the function $\varphi_0(\xi, \theta_0)$. Denote by γ the contour in the complex plane $\theta_0 = \theta_{01} + i\theta_{02}$ consisted from the ray $\theta_{01} = 0$, $\theta_{02} \ge 0$, the segment $0 \le \theta_{01} \le \pi/2$, $\theta_{02} = 0$ and the ray $\theta_{01} = \pi/2$, $\theta_{02} \ge 0$. For the mapping $s = w(k, \theta_0)$ defined by equality (6.4.7), the contour γ is mapped into the real axis $-\infty < s < \infty$ of the plane s for each $k \in (-\infty, \infty)$.

Suppose that the following condition holds:

(D) for each fixed $\xi \in (-\infty, \infty)$ the function $\varphi_0(\theta_0, \xi)$ is analytical in a certain domain D_γ which contains the contour γ inside. Besides, for two certain real values $k = k_1^*$ and $k = k_2^*$, the Fourier transform $\tilde{\varphi}_0(k, \theta_0)$ of the function $\varphi_0(\xi, \theta_0)$ has real or pure imaginary values for all θ_0 belonging to the contour γ. If conditions (A)–(D) hold, then Theorem 6.4.2 holds also in the case when φ_0 depends on θ_0 (the role of the numbers k_1 and k_2 play the numbers k_1^* and k_2^*). In this case, in the real problem 6.4, we define φ_0 as the function of two variables θ_0, ξ.

Using Theorem 5.2.1, we can obtain the analogous results for the analog of the inverse problem 6.4 in the elliptic case. Taking into account Theorems 4.5.1 and 5.5.1, we can transform all the results of Chapter 6 for the problems with the condition of fixed boundary $u(H, \xi, \theta_0) = 0$ instead of $(\partial u/\partial z)(H, \xi, \theta_0) = 0$.

Remark 6.4.1. All the considerations, theorems and formulas of this chapter can be easily transformed for the acoustic case. So, we can formulate the analogous inverse problems on determining the functions $\sigma(x, \theta_0)$, $\overset{+}{\sigma}(x, \theta_0)$, the speed $v(z)$ of the propagation of longitudinal waves and the density $\rho(z)$ in the liquid inhomogeneous layer. The role of equation (6.1.1) or (6.1.5) plays the equation of vertical component of the shift vector (see Remark 3.5.2).

Chapter 7.

Direct and inverse problems for the equations of mixed type

In this chapter we consider the equation of mixed (elliptic-hyperbolic) type of the form

$$K(z)\,\frac{\partial^2 u}{\partial \xi^2} + \frac{\partial^2 u}{\partial z^2} = 0 \tag{7.0.1}$$

in the strip $0 < z < H$, $-\infty < \xi < \infty$, where

$$K(z) < 0 \text{ for } z \in [0, h), \quad K(z) > 0 \text{ for } z \in (h, H], \quad 0 < h < H < \infty.$$

First, in § 7.1, we formulate the direct problem and prove the uniqueness theorem. In § 7.2 we obtain the representation of the direct problem in the particular case $K(h+0) \neq 0$, $K(h-0) \neq 0$, when the coefficient $K(z)$ in equation (7.0.1) is not equal to zero everywhere; therefore, it has a finite jump in the point $z = h$ (in the line $z = h$ the type of equation (7.0.1) is changed). In § 7.3 this representation is illustrated by the example of the Lavrentiev—Bitsadze equation, when $K(z) = \operatorname{sign}(z - h) = \pm 1$.

In § 7.5 we find the representation of solution of the direct problem in the general case, when the coefficient $K(z)$ may vanish from the left or from the right (or both) in the point $z = h$. Moreover, the function $K(z)$ can vanish with an arbitrary real power, generally speaking, different from the right and from the left in the point $z = h$. The proof which establishes this representation of the direct problem in the general case is essentially more complicated than those for the particular case $K(h+0) \neq 0$, $K(h-0) \neq 0$. This proof uses essentially the asymptotic formulas of Langer (1931).

The inverse problems connected with determination of the variable coefficient $K(z)$ in equation (7.0.1) are considered in § 7.4, 7.5, 7.6. The formulations, the way of solution and the proof of uniqueness theorem are given. As the data in these inverse problems we give the solution of the direct problem in one of the boundaries of the strip or in the line $z = h$ (the line passing through which the equation (7.0.1) changes the type). The method of proving of all the uniqueness theorem in this chapter as in the case of considered above problems for hyperbolic and elliptic type equations is spectral-analytic. We reduce these problems to the problem of determination of the regular Sturm—Liouville operator given the two spectra or given the spectral function of regular string with the nonnegative density. This method uses essentially the explicit representation of the solution of the direct problem.

The additional direct and inverse problems for the mixed type equation (7.0.1) are considered in § 7.6.

In § 7.7 the physical interpretation of the direct and inverse problems from this chapter is given. The origin of all these problems is clarified also.

7.1. FORMULATION AND THE UNIQUENESS THEOREM FOR THE DIRECT PROBLEM

Boundary-value problem 7.1. Let in the interval $0 \leq z \leq H$ a real function $K(z)$ be given, continuous in each of the intervals $[0, h]$, $[h, H]$ $(h < H < \infty)$; $K(z) < 0$ for $z \in [0, h)$ and $K(z) > 0$ for $z \in (h, H]$. Each of the conditions

$$K(h+0) \neq K(h-0), \qquad K(h+0) = K(h-0) = 0$$

is possible. Let $\varkappa(\xi)$ be a real continuous function bounded in the line $-\infty < \xi < \infty$, where $\varkappa(\xi)$ either conserves the sign in the whole line, or $\varkappa(\xi) \equiv 0$. Let $\varkappa(\xi)$ be given for $-\infty < \xi < \infty$. Also, the real function $f(\xi)$ for $-\infty < \xi < \infty$ and the numbers h, H are given.

We need to find the function $u(z, \xi)$ satisfying the equation (7.0.1) in the strip $0 < z < h$, $-\infty < \xi < \infty$ and in the strip $h < z < H$, $-\infty < \xi < \infty$; the boundary conditions and the sewing conditions (for $-\infty < \xi < \infty$):

$$\left.\frac{\partial u}{\partial z}\right|_{z=H} = 0, \tag{7.1.1}$$

$$\lim_{z \to h+0} u(z, \xi) = \lim_{z \to h-0} u(z, \xi), \tag{7.1.2}$$

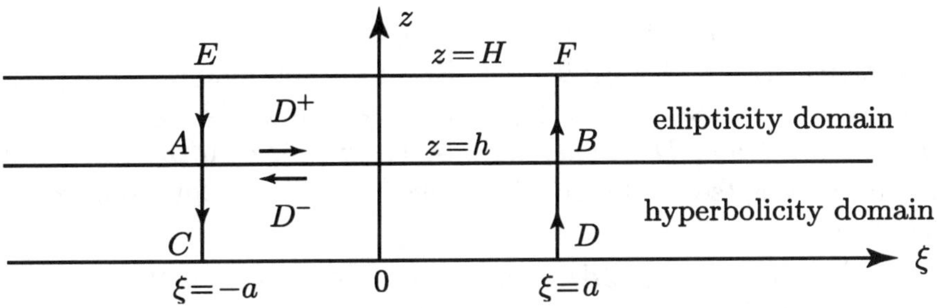

Figure 7.1.

$$\lim_{z\to h+0}\frac{\partial u}{\partial z}=\lim_{z\to h-0}\frac{\partial u}{\partial z}, \tag{7.1.3}$$

$$\left\{\frac{\partial u}{\partial z}-\varkappa\frac{\partial u}{\partial \xi}\right\}\Big|_{z=0}=f(\xi), \tag{7.1.4}$$

and the conditions of smoothness and decreasing: the derivatives u_z, u_ξ, u_{zz}, $u_{\xi\xi}$, $u_{z\xi}$ are continuous in each strip

$$0\le z\le h,\ -\infty<\xi<\infty \quad \text{and} \quad h\le z\le H,\ -\infty<\xi<\infty, \tag{7.1.5}$$

$$u_z,\ u_\xi\to 0 \quad \text{for} \quad \xi\to\pm\infty \tag{7.1.6}$$

uniformly relatively z in the segment $[0,H]$;

$$u(z,\xi)\to 0 \quad \text{for} \quad \xi\to\infty, \quad \text{or} \quad \xi\to-\infty, \quad \text{or} \quad \xi\to\pm\infty \tag{7.1.7}$$

and for certain value $z\in[0,H]$.

We shall call this problem by the boundary-value (direct) problem 7.1.

The following theorem holds.

Theorem 7.1.1. *Let there exist two solutions of direct problem* (7.0.1)–(7.1.6). *Then these solutions may differ only on a constant value. Therefore, the direct problem 7.1 may have the only one solution.*

Proof. Here and below in § 7.1 we shall denote by $u(z,\xi)$ the solution of the homogeneous (for $f(\xi)\equiv 0$) direct problem 7.1.

Let D be the domain bounded by the line intervals $z=0$, $z=H$ and $\xi=-a$, $\xi=a$ $(a>0)$. Denote by D^- the part of D where $0<z<h$ and by D^+ the part of D, where $h<z<H$ (Figure 7.1).

Set

$$P(z,\xi)=-2u_\xi u_z, \qquad Q(z,\xi)=K(z)u_\xi^2-u_z^2. \tag{7.1.8}$$

From condition (7.1.5) it follows that in each of the domains D^- and D^+ the functions $P(z,\xi)$, $Q(z,\xi)$, $\partial P/\partial z$, $\partial Q/\partial\xi$ are continuous. Moreover, taking into account equation (7.0.1), we see that $\partial P/\partial z = \partial Q/\partial\xi$ in each of the domains D^- and D^+. Therefore, by the well-known theorem of mathematical analysis (see, Fikhtengolts, 1968, p. 265), the following equalities hold

$$\int_{S+} P\,\mathrm{d}\xi + Q\,\mathrm{d}z = 0, \qquad \int_{S-} P\,\mathrm{d}\xi + Q\,\mathrm{d}z = 0.$$

The integration is going counter–clockwise along the closed contour S^+ (the boundary of D^+) and S^- (the boundary D^-)[i]. Or, in other words, we have

$$\int_{AB} P\,\mathrm{d}\xi + \int_{EA+BF} Q\,\mathrm{d}z + \int_{FE} P\,\mathrm{d}\xi = 0,$$
$$\int_{BA} P\,\mathrm{d}\xi + \int_{AC+DB} Q\,\mathrm{d}z + \int_{CD} P\,\mathrm{d}\xi = 0. \tag{7.1.9}$$

Summarizing two last equalities and taking into account conditions (7.1.1)–(7.1.4) we obtain

$$\int_{-a}^{a} 2\varkappa u_\xi^2(0,\xi)\,\mathrm{d}\xi = \int_{EC+DF} (Ku_\xi^2 - u_z^2)\,\mathrm{d}z.$$

Therefore, by condition (7.1.6), for each $\varepsilon > 0$, there exists such $\xi = a$, that

$$\left|\int_{-a}^{a} 2\varkappa u_\xi^2(0,\xi)\,\mathrm{d}\xi\right| < \varepsilon.$$

Suppose that $\varkappa(\xi)$ does not change the sign for $-\infty < \xi < \infty$. Then, the last inequality yields

$$u_\xi = 0 \quad \text{for} \quad z = 0, \qquad -\infty < \xi < \infty.$$

Therefore, $u_z = 0$ for $z = 0$, $-\infty < \xi < \infty$, by condition (7.1.4). Note that for $\varkappa \equiv 0$ we have also $u_z(0,\xi) \equiv 0$. Therefore, equalities (7.1.9) take the following form

$$\int_{AB} P\,\mathrm{d}\xi + \int_{EA+BF} Q\,\mathrm{d}z = 0, \quad \int_{BA} P\,\mathrm{d}\xi + \int_{AC+DB} Q\,\mathrm{d}z = 0. \tag{7.1.10}$$

Let in a certain simply connected domain Ω with piecewise smooth boundary σ the functions $K(z)$, $b(z,\xi)$, $v(z,\xi)$ and their derivatives b_z, b_ξ,

---

[i]This theorem and the functions of form (7.1.8) are used in the method of auxiliary functions (Smirnov, 1970, pp. 14, 172).

v_z, v_ξ, v_{zz}, $v_{\xi\xi}$, $v_{z\xi}$ be defined and continuous up to the boundary σ. Then the identity holds

$$2\iint_\Omega bv_\xi(Kv_{\xi\xi}+v_{zz})\,\mathrm{d}\xi\,\mathrm{d}z = \int_\sigma -2bv_\xi v_z\,\mathrm{d}\xi + b(Kv_\xi^2 - v_z^2)\,\mathrm{d}z + \iint_\Omega [v_\xi^2(-Kb_\xi) - 2v_\xi v_z(b_z) + v_z^2(b_\xi)]\,\mathrm{d}\xi\,\mathrm{d}z.^{\text{ii)}}$$

Applying this formula to each of the domains D^+ and D^- and setting $b=\xi$, $v=u(z,\xi)$ we find that

$$\iint_{D^+}(-Ku_\xi^2+u_z^2)\,\mathrm{d}\xi\,\mathrm{d}z + \xi\int_{EA+BF} Q\,\mathrm{d}z + \int_{AB}\xi P\,\mathrm{d}\xi = 0,$$
$$\iint_{D^-}(-Ku_\xi^2+u_z^2)\,\mathrm{d}\xi\,\mathrm{d}z + \xi\int_{AC+DB} Q\,\mathrm{d}z + \int_{BA}\xi P\,\mathrm{d}\xi = 0.$$

Summarizing these equalities, adding with $(+), (-)$ the value $\xi\int_{AB} P\,\mathrm{d}\xi$ and taking into account formulas (7.1.10), we obtain

$$\iint_{D^-+D^+}(-Ku_\xi^2+u_z^2)\,\mathrm{d}\xi\,\mathrm{d}z = 0. \tag{7.1.11}$$

Let

$$Lw = K(z)w_{\xi\xi} + w_{zz}.$$

We set $w = u^2(z,\xi)$, $v\equiv 1$ in the Green formula

$$\iint_\Omega (vLw - wLv)\,\mathrm{d}\xi\,\mathrm{d}z = \int_\sigma \Big[K\Big(v\frac{\partial w}{\partial\xi} - w\frac{\partial v}{\partial\xi}\Big)\Big]\,\mathrm{d}z + \Big[w\frac{\partial v}{\partial z} - v\frac{\partial w}{\partial z}\Big]\,\mathrm{d}\xi$$

for the differential operator L (the integration is going along the boundary σ of the domain Ω counter-clockwise). Then, choosing D^+ and D^- as the domain Ω, we obtain the formulas

$$\iint_{D^+}(Ku_\xi^2+u_z^2)\,\mathrm{d}\xi\,\mathrm{d}z = \int_{AB} -uu_z\,\mathrm{d}\xi + \int_{EA+BF} Kuu_\xi\,\mathrm{d}z,$$
$$\iint_{D^-}(Ku_\xi^2+u_z^2)\,\mathrm{d}\xi\,\mathrm{d}z = \int_{BA} -uu_z\,\mathrm{d}\xi + \int_{AC+DB} Kuu_\xi\,\mathrm{d}z. \tag{7.1.12}$$

Add together these two equalities; we obtain

$$\iint_{D^-+D^+}(Ku_\xi^2+u_z^2)\,\mathrm{d}\xi\,\mathrm{d}z = \int_{EC+DF} Kuu_\xi\,\mathrm{d}z. \tag{7.1.13}$$

---

ii) This is a particular form of the *abc* method (see, Smirnov, 1970, p. 177)

Adding together equalities (7.1.11) and (7.1.13), we obtain, as a result

$$2 \iint_{D^- + D^+} u_z^2 \, \mathrm{d}\xi \, \mathrm{d}z = \int_{EC+DF} K u u_\xi \, \mathrm{d}z.$$

By condition (7.1.6), for sufficiently large a and for each $\varepsilon > 0$, the integral in the right–hand side is less than ε by the modulus; therefore,

$$u_z \equiv 0 \quad \text{for} \quad 0 \le z \le H, \quad -\infty < \xi < \infty.$$

In this case formulas (7.1.12) yield

$$u_\xi \equiv 0 \quad \text{for} \quad 0 \le z \le H, \quad -\infty < \xi < \infty.$$

The theorem is proved. □

If we change the domains where $K > 0$ and $K < 0$, then all the statements and formulas of § 1 hold true if we change D^- by D^+ and conversely.

7.2. THE REPRESENTATION OF SOLUTION OF THE DIRECT PROBLEM 7.1. THE CASE OF $K(h+0) \neq 0$, $K(h-0) \neq 0$

Theorem 7.2.1. *Let in conditions of the direct problem 7.1*

$$\varkappa(\xi) \equiv \text{const} = \varkappa \in (-\infty, \infty)$$

where $\varkappa \neq 0$ and $K(h+0) \neq 0$, $K(h-0) \neq 0$ (in this case the function $K(z)$ has a finite jump passing through the point $z = h$).

Introduce the positive functions $\bar{V}(z)$ and $\overset{+}{V}(z)$ as follows:

$$K(z) = \begin{cases} -1/\bar{V}^2(z), & 0 \le z < h, \\ 1/\overset{+}{V}{}^2(z), & h < z \le H \end{cases}$$

and set

$$f(\xi) = \frac{\mathrm{d}\varphi}{\mathrm{d}\xi} \qquad (\varphi(0) = 0).$$

Let the conditions hold:

(A) the derivative $\mathrm{d}^2K/\mathrm{d}z^2$ is continuous in each of the segments $[0, h]$ and $[h, H]$;

(B) *the function $\varphi(\xi)$ and its derivatives up to the fourth order are continuous and absolutely integrable in the line $-\infty < \xi < \infty$;*

(C) *the function $\xi\varphi(\xi)$ are absolutely integrable in the line $-\infty < \xi < \infty$.*

Then the solution $u(z,\xi)$ of the direct problem 7.1 exists, is unique and has the form

$$u(z,\xi) = \int_{-\infty}^{\infty} U(z,k)e^{ik\xi}\,\mathrm{d}k, \qquad 0 \le z \le H, \quad -\infty < \xi < \infty, \tag{7.2.1}$$

where the integration holds along the real axis of the complex plane k;

$$\begin{gathered} U(z,k) = \bar{V}^{1/2}(z)W_h(x,k), \qquad x = \int_0^z \frac{\mathrm{d}z}{\bar{V}(z)}, \\ 0 \le z \le h, \qquad 0 \le x \le x_h = x(h), \end{gathered} \tag{7.2.2}$$

$$\begin{gathered} U(z,k) = \overset{+}{V}{}^{1/2}(z)W_e(x_1,k), \qquad x_1 = \int_h^z \frac{\mathrm{d}z}{\overset{+}{V}(z)}, \\ h \le z \le H, \qquad 0 \le x_1 \le x_H = x_1(H), \\ W_e(x_1,k) = \frac{\gamma ik\tilde{\varphi}(k)\{\Phi_e(ik)\psi_e(x_1,ik) - \Psi_e(ik)\varphi_e(x_1,ik)\}}{A_1(k)+ik\beta A_2(k)}, \\ W_h(x,k) = \frac{\gamma ik\tilde{\varphi}(k)\{A_1(k)\psi_h(x,k) - A_2(k)\varphi_h(x,k)\}}{A_1(k)+ik\beta A_2(k)}, \\ 0 \le x_1 \le x_H, \qquad 0 \le x \le x_h, \qquad -\infty < k < \infty. \end{gathered} \tag{7.2.3}$$

Here

$$A_1(k) = \delta_1\Phi_e(ik)\varphi_h(x_h,k) + \delta_2\Psi_e(ik)\Phi_h(k), \tag{7.2.4}$$
$$A_2(k) = \delta_1\Phi_e(ik)\psi_h(x_h,k) + \delta_2\Psi_e(ik)\Psi_h(k), \tag{7.2.5}$$

$$\begin{gathered} \Phi_e(s) = \varphi_e'(x_H,s) - \overset{+}{\alpha}\,\varphi_e(x_H,s), \qquad \Psi_e(s) = \psi_e'(x_H,s) - \overset{+}{\alpha}\,\psi_e(x_H,s), \\ \Phi_h(k) = \varphi_h'(x_h,k) - \bar{\alpha}\varphi_h(x_h,k), \qquad \Psi_h(k) = \psi_h'(x_h,k) - \bar{\alpha}\psi_h(x_h,k), \end{gathered} \tag{7.2.6}$$

The functions $\varphi_e(x_1,s)$ and $\psi_e(x_1,s)$ solve the equation

$$-y'' + \overset{+}{q}(x_1)y = s^2 y, \qquad 0 \le x_1 \le x_H \tag{7.2.7}$$

with the function

$$\overset{+}{q}(x_1) = \overset{+}{v}^{1/2}(x_1)\{\overset{+}{v}^{-1/2}(x_1)\}'' \qquad (\overset{+}{v}(x_1) = \overset{+}{V}(z), \quad x_1 = x_1(z)) \quad (7.2.8)$$

respectively, for the initial conditions

$$\varphi_e(0,s) = 1, \qquad \varphi_e'(0,s) = \overset{+}{\alpha}_0, \quad (7.2.9)$$
$$\psi_e(0,s) = 0, \qquad \psi_e'(0,s) = 1. \quad (7.2.10)$$

The functions $\varphi_h(x,k)$ and $\psi_h(x,k)$ solve the equation

$$-y'' + \bar{q}(x)y = k^2 y, \qquad 0 \le x \le x_h \quad (7.2.11)$$

with the function

$$\bar{q}(x) = \bar{v}^{1/2}(x)\{\bar{v}^{-1/2}(x)\}'' \qquad (\bar{v}(x) = \bar{V}(z), \quad x = x(z)) \quad (7.2.12)$$

respectively, for the initial conditions

$$\varphi_h(0,k) = 1, \qquad \varphi_h'(0,k) = \bar{\alpha}_0, \quad (7.2.13)$$
$$\psi_h(0,k) = 0, \qquad \psi_h'(0,k) = 1; \quad (7.2.14)$$

the function $\tilde{\varphi}(k)$ is the Fourier transform of the function $\varphi(\xi)$:

$$\varphi(\xi) = \int_{-\infty}^{\infty} \tilde{\varphi}(k) e^{ik\xi}\, dk.$$

The numbers $\overset{+}{\alpha}_0$, $\overset{+}{\alpha}$, $\bar{\alpha}_0$, $\bar{\alpha}$, β, γ, δ_1, δ_2 are expressed by the formulas

$$\overset{+}{\alpha}_0 = -\frac{1}{2}\frac{\overset{+}{v}'(0)}{\overset{+}{v}(0)}, \quad \overset{+}{\alpha} = -\frac{1}{2}\frac{\overset{+}{v}'(x_H)}{\overset{+}{v}(x_H)}, \quad \bar{\alpha}_0 = -\frac{1}{2}\frac{\bar{v}'(0)}{\bar{v}(0)}, \quad \bar{\alpha} = -\frac{1}{2}\frac{\bar{v}'(x_h)}{\bar{v}(x_h)},$$
$$\beta = \varkappa\bar{v}(0), \quad \gamma = \bar{v}^{1/2}(0), \quad \delta_1 = \bar{v}^{1/2}(x_h)\overset{+}{v}^{-1/2}(0), \quad \delta_2 = 1/\delta_1. \quad (7.2.15)$$

The prime here and further will denote the differentiation with respect to x and x_1 respectively.

Thus, these formulas give the explicit expression of solution of the direct problem 7.1 (under the conditions of Theorem 7.2.1) via the fundamental systems of solutions of two Sturm—Liouville equations. The coefficients of the first equation (7.2.7) is constructed by the values of the function $K(z)$ in the domain where equation (7.0.1) is elliptic. The solutions φ, ψ of the first equation (7.2.7) we mark by the sign e, from the word elliptic; and the functions Φ_e, Ψ_e are constructed by the functions φ_e, ψ_e. The coefficient of the second equation (7.2.11) is constructed by the values of the function $K(z)$ in the domain where the equation is hyperbolic. The solutions φ, ψ of the second equation of (7.2.11) we mark by the sign h (from the word hyperbolic). Analogously, the functions Φ_h, Ψ_h are constructed by the solutions φ_h and ψ_h.

Proof of Theorem 7.2.1. Note that the functions $W_e(x_1, k)$ and $W_h(x, k)$ defined by formulas (7.2.3)–(7.2.15) satisfy the following equations and conditions:

$$\frac{\mathrm{d}^2 W_e}{\mathrm{d}x_1^2} + \{s^2 - \overset{+}{q}(x_1)\}W_e = 0, \qquad 0 \le x_1 \le x_H, \quad s = ik, \tag{7.2.16}$$

$$\frac{\mathrm{d}^2 W_h}{\mathrm{d}x^2} + \{k^2 - \bar{q}(x)\}W_h = 0, \quad 0 \le x \le x_h, \tag{7.2.17}$$

$$\Big\{\frac{\mathrm{d}W_e}{\mathrm{d}x_1} - \overset{+}{\alpha} W_e\Big\}\Big|_{x_1 = x_H} = 0, \tag{7.2.18}$$

$$W_e(0, k) = \delta_1 W_h(x_h, k), \tag{7.2.19}$$

$$\Big\{\frac{\mathrm{d}W_e}{\mathrm{d}x_1} - \overset{+}{\alpha}_0 W_e\Big\}\Big|_{x_1=0} = \delta_2\Big\{\frac{\mathrm{d}W_h}{\mathrm{d}x} - \bar{\alpha} W_h\Big\}\Big|_{x=x_h}, \tag{7.2.20}$$

$$\Big\{\frac{\mathrm{d}W_h}{\mathrm{d}x} - (\bar{\alpha}_0 + ik\beta)W_h\Big\}\Big|_{x=0} = \gamma ik\tilde{\varphi}(k). \tag{7.2.21}$$

Equations (7.2.16), (7.2.17) and conditions (7.2.18)–(7.2.21) are given in the real axis $-\infty < k < \infty$ of the complex plane k (therefore, equation (7.2.16) is given in the imaginary axis of the complex plane $s = ik$). The functions $\overset{+}{q}(x_1)$, $\bar{q}(x)$ and the numbers $\overset{+}{\alpha}_0$, $\overset{+}{\alpha}$, $\bar{\alpha}_0$, $\bar{\alpha}$, β, γ, δ_1, δ_2 are real and expressed by formulas (7.2.8), (7.2.12), (7.2.15).

Really, taking into account the condition (A), we see that the functions $\overset{+}{q}(x_1)$ and $\bar{q}(x)$ are continuous in the segments $[0, x_H]$, $[0, x_h]$ respectively. Therefore, equations (7.2.7) and (7.2.16) have the solutions $\varphi_e(x_1, s)$, $\psi_e(x_1, s)$ respectively for initial conditions (7.2.9), (7.2.10). Equations (7.2.11) and (7.2.17) have the solutions $\varphi_h(x, k)$ and $\psi_h(x, k)$ respectively with initial conditions (7.2.13), (7.2.14) with the following properties (Faddeev, 1959; Levitan and Sargsyan, 1970). For each fixed value $x_1 \in [0, x_H]$, the values $\varphi_e(x_1, s)$ and $\psi_e(x_1, s)$ are the entire analytic functions with respect to s. They are continuous with respect to (x_1, s) for $x_1 \in [0, x_H]$ and

$-\infty < s < \infty$ and form the fundamental system of solutions of equation (7.2.16); their Wronskian is as follows:

$$\psi_e'(x_1, s)\varphi_e(x_1, s) - \psi_e(x_1, s)\varphi_e'(x_1, s) \equiv 1. \tag{7.2.22}$$

The analogous properties have the solutions $\varphi_h(x, k)$ and $\psi_h(x, k)$. We represent the solution $\{W_e(x_1, k),\ W_h(x, k)\}$ in the form

$$\begin{aligned} W_e(x_1, k) &= C_1(k)\varphi_e(x_1, s) + C_2(k)\psi_e(x_1, s), \\ W_h(x, k) &= D_1(k)\varphi_h(x, k) + D_2(k)\psi_h(x, k) \end{aligned}$$

and derive the coefficients $C_1(k), C_2(k), D_1(k), D_2(k)$ from conditions (7.2.18)–(7.2.21). As a result we obtain for W_e, W_h the formulas (7.2.3)–(7.2.6).

The numbers $\overset{+}{\alpha}, \bar{\alpha}, \beta, \delta_1, \delta_2$ are real. The functions $\varphi_e(x_1, s)$, $\psi_e(x_1, s)$ $(s = ik)$, $\varphi_h(x, k)$, $\psi_h(x, k)$ are real also for $-\infty < k < \infty$.

Therefore, the functions $A_1(k)$ and $A_2(k)$ are real for $-\infty < k < \infty$. Besides, the following statement holds.

Lemma 7.2.1. *The functions $A_1(k)$ and $A_2(k)$ defined by formulas* (7.2.4)–(7.2.6) *do not vanish simultaneously for each $k \in (-\infty, \infty)$.*

Proof. Suppose, for a certain k, we have

$$A_1(k) = A_2(k) = 0.$$

Then, the functions $\psi_h(x_h, k)A_1(k)$ and $\varphi_h(x_h, k)A_2(k)$ are equal to zero also. Therefore, for such k, we have

$$\psi_h(x_h, k)A_1(k) = \varphi_h(x_h, k)A_2(k). \tag{7.2.23}$$

Substituting in this equality the expressions (7.2.4)–(7.2.6) for $A_1(k)$ and $A_2(k)$ we obtain the equality for $s = ik$:

$$\delta_2\Psi_e(s)\{\varphi_h(x_h, k)\psi_h'(x_h, k) - \psi_h(x_h, k)\varphi_h'(x_h, k)\} = 0. \tag{7.2.24}$$

Since the Wronskian is equal to unity:

$$\varphi_h(x, k)\psi_h'(x, k) - \psi_h(x, k)\varphi_h'(x, k) \equiv 1,$$

the equality (7.2.24) yields

$$\delta_2\Psi_e(s) = 0.$$

We have $\delta_2 \neq 0$; therefore, the last equality is impossible if $s = ik$ is not a zero of the function $\Psi_e(s)$. So, the values of the functions $A_1(k)$, $A_2(k)$ are not equal to zero simultaneously if $s = ik$ is not a zero of the function $\Psi_e(s)$.

Let $s = ik$ be a zero of the function $\Psi_e(s)$. Then, taking into account formulas (7.2.4), (7.2.5), we have in this point k:

$$A_1(k) = \delta_1 \Phi_e(s) \varphi_h(x_h, k), \qquad A_2(k) = \delta_1 \Phi_e(s) \psi_h(x_h, k).$$

The supposition $A_1(k) = A_2(k) = 0$ yields also

$$\Psi_h(k) A_1(k) = \Phi_h(k) A_2(k),$$

whence it follows that

$$\delta_1 \Phi_e(s) = 0.$$

As $\delta_1 \neq 0$, then $\Phi_e(s) = 0$. Therefore, in the point k that we take, the two equalities hold simultaneously

$$\Psi_e(s) = 0, \qquad \Phi_e(s) = 0.$$

This, however, contradicts to the identity

$$\varphi_e(x_H, s)\Psi_e(s) - \psi_e(x_H, s)\Phi_e(s) \equiv \varphi_e(x_H, s)\psi_e'(x_H, s) - \psi_e(x_H, s)\varphi_e'(x_H, s) \equiv 1.$$

Hence the lemma follows. □

From Lemma 7.2.1 it follows that function

$$A(k) = A_1(k) + ik\beta A_2(k)$$

does not vanish for $\varkappa \neq 0$, $-\infty < k < \infty$, $k \neq 0$. The explicit formulas

$$\varphi_e(x_1, 0) = \overset{+}{v}{}^{1/2}(0)\, \overset{+}{v}{}^{-1/2}(x_1), \qquad \varphi_h(x, 0) = \bar{v}^{1/2}(0)\bar{v}^{-1/2}(x), \tag{7.2.25}$$

$$\psi_e(x_1, 0) = \overset{+}{v}{}^{1/2}(0)\, \overset{+}{v}{}^{-1/2}(x_1) \int_0^{x_1} \overset{+}{v}(x_1)\, \mathrm{d}x_1, \tag{7.2.26}$$

$$\psi_h(x, 0) = \bar{v}^{-1/2}(0)\bar{v}^{-1/2}(x) \int_0^{x} \bar{v}(x)\, \mathrm{d}x \tag{7.2.27}$$

yield $A_1(0) = 0$, $A_2(0) \neq 0$, so, the point $k = 0$ is the zero of the first order of the function $A(k)$. Therefore, the functions $W_e(x_1, k)$ and $W_h(x, k)$ are continuous for $-\infty < k < \infty$.

Moreover, from the condition (C) and the analytic properties of the functions $\varphi_e(x_1, s)$, $\psi_e(x_1, s)$, $\varphi_h(x, k)$, $\psi_h(x, k)$ it follows that the functions $W_e^{(n)}(x_1, k)$ and $W_h^{(n)}(x, k)$ $(n = 0, 1, 2)$ are differentiable with respect to k in the real axis $-\infty < k < \infty$ (except may be the point $k = 0$) for each fixed $x_1 \in [0, x_H]$, $x \in [0, x_h]$. Further, using the asymptotic properties of the functions $\varphi_e(x_1, s)$, $\psi_e(x_1, s)$, $\varphi_h(x, k)$, $\psi_h(x, k)$ (see Faddeev, 1959; Levitan and Sargsyan, 1970, p. 17), we establish the following lemma.

Lemma 7.2.2. *There exists such a number $k_0 > 0$, that for $|k| \geq k_0$, $n = 0, 1, 2$, the estimates*

$$|W_e^{(n)}(x_1, k)| < C_e|k|^n|\tilde{\varphi}(k)|, \qquad |W_h^{(n)}(x, k)| < C_h|k|^n|\tilde{\varphi}(k)|$$

hold, where the constants C_e, C_h are independent of $x_1 \in [0, x_H]$, $x \in [0, x_h]$[iii]*.*

Proof. Let $\varphi(x, k)$ and $\psi(x, k)$ solve the equation

$$-y'' + q(x)y = k^2 y, \qquad 0 \leq x \leq a \tag{7.2.28}$$

respectively for the initial data

$$\varphi(0, k) = 1, \quad \varphi'(0, k) = \alpha, \qquad \psi(0, k) = 0, \quad \psi'(0, k) = 1,$$

where $q(x)$ is a continuous real function in the interval $[0, a]$; α is a real number. As is known (Faddeev, 1959; Levitan and Sargsyan, 1970, p. 17), the functions $\varphi(x, k)$, $\psi(x, k)$ have the following asymptotic properties for $|k| \to \infty$. There exists such $k_0 > 0$ that for $|k| > k_0$ $(k = k_1 + ik_2)$ the following asymptotic formulas hold:

$$\begin{aligned}
\varphi(x, k) &= \cos kx + \alpha \sin(kx)/k + O(e^{|k_2|x}/|k|), \\
\psi(x, k) &= \sin(kx)/k + O(e^{|k_2|x}/|k|^2), \\
\varphi'(x, k) &= -k \sin kx + \alpha \cos kx + O(e^{|k_2|x}/k), \\
\psi'(x, k) &= \cos kx + O(e^{|k_2|x}/|k|), \\
\varphi''(x, k) &= -k^2\varphi + q(x)\varphi = -k^2 \cos kx - \alpha k \sin kx + O(|k|e^{|k_2|x}), \\
\psi''(x, k) &= -k^2\psi + q(x)\psi = -k \sin kx + O(e^{|k_2|x}).
\end{aligned}$$

[iii] Here, before and further we use the notation $f^{(n)}(x, k)$ for the derivative of the function of the order n with respect to the variable x.

Since

$$\sin sx = (e^{isx} - e^{-isx})/(2i), \qquad \cos sx = (e^{isx} + e^{-isx})/2,$$

then, for $s = ik$, we have

$$\sin(ikx) = (e^{-kx} - e^{kx})/(2i), \qquad \cos(ikx) = (e^{-kx} + e^{kx})/2.$$

These formulas yield the following estimates for the functions $A_1(k)$, $A_2(k)$ defined by formulas (7.2.4), (7.2.5) in the real axis k

$$\begin{aligned} A_1(k) &= (k/2)\{e^{-kx_H}[-\delta_1 \cos kx_h - \delta_2 \sin kx_h] \\ &\qquad + e^{kx_H}[\delta_1 \cos kx_h - \delta_2 \sin kx_h]\} + O(e^{|k|x_H}), \\ A_2(k) &= (1/2)\{e^{-kx_H}[-\delta_1 \sin kx_h + \delta_2 \cos kx_h] \\ &\qquad + e^{kx_H}[\delta_1 \sin kx_h + \delta_2 \cos kx_h]\} + O(e^{|k|x_H}/|k|). \end{aligned}$$

Since

$$|A(k)|^2 = \{A_1(k)\}^2 + \beta^2 k^2 \{A_2(k)\}^2,$$

then, from the estimates that we obtained for $A_1(k)$, $A_2(k)$, we have that the modulus of $A(k)$ increases as $\{|k|e^{|k|x_H}\}$ for $|k| \to \infty$. Therefore, for $|k|$ sufficiently large, the function $A(k)$ satisfies the estimate

$$|A(k)| \geq Q|k|e^{x_H|k|}, \tag{7.2.29}$$

where Q is a certain positive constant.

We obtain now the estimate for the function

$$\begin{aligned} A_3(x_1, k) &\equiv \Phi_e(ik)\psi_e(x_1, ik) - \Psi_e(ik)\varphi_e(x_1, ik) \\ &= \varphi_e'(x_H, s)\psi_e(x_1, s) - \psi_e'(x_H, s)\varphi_e(x_1, s) \\ &\quad - \overset{+}{\alpha}\{\varphi_e(x_H, s)\psi_e(x_1, s) - \psi_e(x_H, s)\varphi_e(x_1, s)\} \end{aligned} \tag{7.2.30}$$

which stands in the numerator of the expression for the function $W_e(x_1, k)$ in formula (7.2.3).

Return to equation (7.2.28), where we set $a = x_H$ and $q(x)$ is a continuous real function in the segment $[0, x_H]$. Introduce the function $q_H(x)$ by the formula

$$q_H(x) = \begin{cases} q(x), & 0 \leq x \leq x_H \\ 0, & x > x_H. \end{cases}$$

Let $f(x,k)$ and $\tilde{f}(x,k)$ be two solutions of equation

$$-y'' + q_H(x)y = k^2 y, \qquad 0 \le x < \infty, \tag{7.2.31}$$

defined by the conditions

$$\lim_{x\to\infty} e^{-ikx} f(x,k) = 1, \quad \lim_{x\to\infty} e^{ikx}\tilde{f}(x,k) = 1. \tag{7.2.32}$$

Equation (7.2.31) with conditions (7.2.32) is equivalent to the integral Volterra equations of the second kind (see Faddeev, 1959; Levitan and Sargsyan, 1970)

$$f(x,k) = e^{ikx} + \int_x^{x_H} \frac{\sin k(t-x)}{k} q(t) f(t,k)\,\mathrm{d}t,$$
$$\tilde{f}(x,k) = e^{-ikx} + \int_x^{x_H} \frac{\sin k(t-x)}{k} q(t) \tilde{f}(t,k)\,\mathrm{d}t.$$

We had considered the solutions $f(x,k)$, $\tilde{f}(x,k)$ in Chapter 1, where they were defined in the half-planes $\operatorname{Im} k \ge 0$ and $\operatorname{Im} k \le 0$ respectively. In our case, since the function $q_H(x)$ is finite, the solutions $f(x,k)$ and $\tilde{f}(x,k)$ are defined and continuous with respect to (x,k) in the whole complex plane k for $0 \le x < \infty$. For each fixed value $x \in [0,\infty)$, the functions $f(x,k)$ and $\tilde{f}(x,k)$ are entire analytic functions with respect to k, and we have

$$\tilde{f}(x,k) = f(x,-k).$$

These functions have the property

$$f(x,k) = e^{ikx}, \qquad \tilde{f}(x,k) = e^{-ikx} \qquad \text{for} \quad x \ge x_H. \tag{7.2.33}$$

For $|k| \to \infty$ these functions are as follows:

$$f_x^{(n)}(x,k) = (ik)^n e^{ikx}[1 + O(1/k)], \qquad n = 0,1,2$$

uniformly relatively $x \in [0, x_H]$.

All these properties (analytic and asymptotic) of the functions $f(x,k)$ and $\tilde{f}(x,k)$ in the case of finite coefficients $q(x)$ follows, for example from the paper of Naymark (1954) (the condition of Theorem 2.6.1 of this paper holds for each $\varepsilon > 0$ for finite functions $q(x)$).

Note that the following identities hold

$$\varphi(x,k) = \{[f'(0,k) - \alpha f(0,k)]\tilde{f}(x,k) - [\tilde{f}'(0,k) - \alpha\tilde{f}(0,k)]f(x,k)\}/(2ik),$$
$$\psi(x,k) = \{\tilde{f}(0,k)f(x,k) - f(0,k)\tilde{f}(x,k)\}/(2ik),$$

which yield the explicit expressions of the solutions $\varphi(x,k)$, $\psi(x,k)$ of equation (7.2.28) or (7.2.31) via the solutions $f(x,k)$, $\tilde{f}(x,k)$. Taking these formulas into account, we obtain

$$\varphi(x,k)\psi'(x_0,k)-\psi(x,k)\varphi'(x_0,k)=\frac{\tilde{f}(x,k)f'(x_0,k)-f(x,k)\tilde{f}'(x_0,k)}{2ik},$$

$$\psi(x,k)\varphi(x_0,k)-\varphi(x,k)\psi(x_0,k)=\frac{f(x,k)\tilde{f}(x_0,k)-\tilde{f}(x,k)f(x_0,k)}{2ik},$$

where x and x_0 are arbitrary points from the ray $0\le x<\infty$. This and equalities (7.2.33) yield (set $x_0=x_H$)

$$\begin{aligned}A_5(x,k)&\equiv\psi'(x_H,k)\varphi(x,k)-\varphi'(x_H,k)\psi(x,k)\\&=\{\tilde{f}(x,k)e^{ikx_H}+f(x,k)e^{-ikx_H}\}/2,\\A_6(x,k)&\equiv\varphi(x_H,k)\psi(x,k)-\psi(x_H,k)\varphi(x,k)\\&=\{f(x,k)e^{-ikx_H}-\tilde{f}(x,k)e^{ikx_H}\}/(2ik).\end{aligned}$$

From asymptotic properties of the function $f(x,k)$ it follows that for $0\le x\le x_H$, $|k|\to\infty$ the function $f^{(n)}(x,k)e^{-ikx_H}$ $(n=0,1,2)$ increases in the imaginary axis no faster than $|k|^n e^{|k|x_H}$. The analogous estimates hold for the function $\tilde{f}^{(n)}(x,k)e^{ikx_H}$ $(n=0,1,2)$. Therefore, the function $A_5^{(n)}(x,k)$ increases in the imaginary axis no faster than $|k|^n e^{|k|x_H}$, and the function $A_6^{(n)}(x,k)$ no faster than $|k|^{n-1}e^{|k|x_H}$ $(n=0,1,2)$.

Applying these estimates to the function $A_3(x_1,k)$ defined by formula (7.2.30) and taking into account that

$$A_3(x_1,k)=-A_5(x_1,s)-\overset{+}{\alpha}A_6(x_1,s)\qquad(s=ik)$$

(the role of x plays x_1 and the role of φ,ψ play φ_e,ψ_e), we find that the function $A_3^{(n)}(x_1,k)$ increases for $|k|\to\infty$ no faster than $|k|^n e^{x_H|k|}$ uniformly relatively $x_1\in[0,x_H]$. This means that for $|k|$ sufficiently large, we have

$$|A_3^{(n)}(x_1,k)|\le P_n|k|^n e^{|k|x_H}\qquad(n=0,1,2),\tag{7.2.34}$$

where the positive constants P_n are independent of $x_1\in[0,x_H]$, $k\in(-\infty,\infty)$. Since

$$W_e^{(n)}(x_1,k)=\gamma ik\tilde{\varphi}(k)A_3^{(n)}(x_1,k)/A(k),$$

then, from the estimates (7.2.29), (7.2.34) it follows that for $|k|$ sufficiently large $(n=0,1,2)$:

$$|W_e^{(n)}(x_1,k)|\le C_e|k|^n|\tilde{\varphi}(k)|,\qquad C_e\ge\gamma\max_n P_n/Q$$

for $x_1\in[0,x_H]$ in the real axis $-\infty<k<\infty$ which was to be proved.

Applying the asymptotic formulas for the functions $\varphi_h(x,k)$, $\psi_h(x,k)$ and taking into account the estimates

$$|A_1(k)| \le a_1|k|e^{|k|x_H}, \qquad |A_2(x)| \le a_2 e^{|k|x_H},$$

where a_1, a_2 are certain positive constants, we obtain for the function

$$A_4(x,k) = A_1(k)\psi_h(x,k) - A_2(k)\varphi_h(x,k)$$

which stands in the numerator of expression (7.2.3) the estimate

$$|A_4^{(n)}(x,k)| \le R_n|k|^n e^{|k|x_H} \qquad (n = 0,1,2)$$

for $|k|$ sufficiently large, $x \in [0, x_h]$, in the real axis $-\infty < k < \infty$. Hence the lemma estimate for the function $W_h(x,k)$ follows. Lemma 7.2.2 is proved. □

Taking into account the estimates of Lemma 7.2.2, the condition (B) and the theorem connection between the smoothness of the function $\varphi(\xi)$ and the decreasing of its Fourier transform $\tilde{\varphi}(k)$ (see Shilov, 1970, p. 277), we obtain, that such $k_1 > 0$ exists that for $|k| \ge k_1$, $n = 0,1,2$ the estimates

$$\begin{aligned} &|k^n W_e(x_1,k)| < K_e/k^{4-n}, \quad |W_e^{(n)}(x_1,k)| < K_e/k^{4-n}, \\ &|k^n W_h(x,k)| < K_h/k^{4-n}, \qquad |W_h^{(n)}(x,k)| < K_h/k^{4-n} \end{aligned} \tag{7.2.35}$$

hold. The positive constants K_e, K_h here are independent of $x_1 \in [0, x_H]$, $x \in [0, x_h]$. These estimates yield that each of the functions $k^n W_e(x_1,k)$, $W_e^{(n)}(x_1,k)$, $k^n W_h(x,k)$, $W_h^{(n)}(x,k)$ $(n = 0,1,2)$ is absolutely integrable in the line $-\infty < k < \infty$; therefore, (see Titchmarsh, 1948, p. 58) it can be represented by the Fourier integral (as a complex-valued function dependent on the real variable k) for each $x_1 \in [0, x_H]$, $x \in [0, x_h]$. Besides, the integral (7.2.1) admits the differentiation twice with respect to z and with respect to ξ under the integral sign. Since $u(z,\xi)$ is the Fourier transform of the function $U(z,k)$ which can be represented as the Fourier integral then the integral in equality (7.2.1) is the Fourier integral (just as the integrals for the derivatives $u_z, u_\xi, u_{zz}, u_{\xi\xi}, u_{z\xi}$).

Finally, the relations hold

$$W_e^{(n)}(x_1,-k) = \bar{W}_e^{(n)}(x_1,k), \qquad W_h^{(n)}(x,-k) = \bar{W}_h^{(n)}(x,k) \qquad (n = 0,1,2)$$

so, all these integrals are real.

From the above properties of the functions $W_e(x_1,k)$ and $W_h(x,k)$ follow that the function $u(z,\xi)$ defined by formulas (7.2.1)–(7.2.15) satisfies to all the conditions of the direct problem 7.1. Theorem 7.2.1 is proved. □

We have shown that under the conditions of Theorem 7.2.1 the solution of the direct problem 7.1 exists. From estimates (7.2.35) and the Riemann—Lebesgue theorem it follows that $u(z,\xi)\to 0$ uniformly relatively z in the interval $[0,H]$. Hence, by Theorem 7.1.1, we obtain that the solution of the direct problem 7.1 (under the conditions of Theorem 7.2.1) is unique and coincides with the solution constructed before.

Stability estimate. From inequalities of Lemma 7.2.2 it follows that if $u^{(1)}(z,\xi)$ and $u^{(2)}(z,\xi)$ are the solutions of the problem 7.1 with the functions

$$f^{(1)}(\xi)=\frac{\mathrm{d}\varphi^{(1)}}{\mathrm{d}\xi}\qquad\text{and}\qquad f^{2}(\xi)=\frac{\mathrm{d}\varphi^{(2)}}{\mathrm{d}\xi}$$

given in condition (7.1.4) ($\varphi^{(1)}(0)=\varphi^{(2)}(0)=0$ in this case), and $\varkappa\neq 0$, then, for $0\le z\le H$, $-\infty<\xi<\infty$, the inequalities

$$|u^{(1)}(z,\xi)-u^{(2)}(z,\xi)|\le C\int_{-\infty}^{\infty}|\tilde\varphi^{(1)}(k)-\tilde\varphi^{(2)}(k)|\,\mathrm{d}k,$$

$$\left|\frac{\partial^n u^{(1)}(z,\xi)}{\partial z^n}-\frac{\partial^n u^{(2)}(z,\xi)}{\partial z^n}\right|\le C\int_{-\infty}^{\infty}|k|^n|\tilde\varphi^{(1)}(k)-\tilde\varphi^{(2)}(k)|\,\mathrm{d}k,$$

$$\left|\frac{\partial^n u^{(1)}(z,\xi)}{\partial \xi^n}-\frac{\partial^n u^{(2)}(z,\xi)}{\partial \xi^n}\right|\le C\int_{-\infty}^{\infty}|k|^n|\tilde\varphi^{(1)}(k)-\tilde\varphi^{(2)}(k)|\,\mathrm{d}k \qquad (7.2.36)$$

hold; where $n=1,2$; C is a certain positive constant, $\tilde\varphi^{i}(k)$ is the Fourier transform of the function $\varphi^{i}(\xi)$, $i=1,2$. The analogous inequality holds for mixed derivatives with the parameter $n=2$.

7.3. THE CASE OF LAVRENTIEV—BITSADZE EQUATION. THE FORMULAS FOR SOLUTION OF THE DIRECT PROBLEM 7.1

We shall demonstrate now the character of formulas (7.2.1)–(7.2.15) obtained in § 7.2 for the solution $u(z,\xi)$ of the direct problem 7.1. For this purpose we consider the particular case of equation (7.0.1), namely, the Lavrentiev—Bitsadze equation. In this case, the coefficient $K(z)$ in the equation (7.0.1) is as follows:

$$K(z)=\begin{cases}1, & \text{for}\quad h<z\le H,\\ -1, & \text{for}\quad 0\le z<h.\end{cases}\qquad \text{(L–B)}$$

We shall call this case by the (L–B) case.

Theorem 7.2.1 has the following formulation in this (L–B) case.

Theorem 7.3.1. *Let $\varkappa \neq 0$ and the conditions (B), (C) for the function $\varphi(\xi)$ from Theorem 7.2.1 hold. Then the solution of the direct problem 7.1 in the (L–B) case exists and has the representation*

$$u(z,\xi) = \int_{-\infty}^{\infty} U(z,k)e^{ikz}\,\mathrm{d}k, \qquad 0 \le z \le H, \quad -\infty < \xi < \infty.$$

The integration goes along the real axis of the complex plane k;

$$U(z,k) = \begin{cases} U_e(z,k), & h \le z \le H, \\ U_h(z,k), & 0 \le z \le h, \end{cases} \tag{7.3.1}$$

$$U_e(z,k) = -\tilde{\varphi}(k)\,\frac{ik\cosh\{k(H-z)\}}{A_1(k) + ik\varkappa A_2(k)}, \tag{7.3.2}$$

$$U_h(z,k) = \tilde{\varphi}(k)\,\frac{ik\{\sin(z-h)k\sinh(h_1k) - \cos(z-h)k\cosh(h_1k)\}}{A_1(k) + ik\varkappa A_2(k)}.$$

Here $h_1 = H - h$,

$$\begin{aligned} A_1(k) &= k\{\cos(hk)\sinh(h_1k) - \sin(hk)\cosh(h_1k)\}, \\ A_2(k) &= \sin(hk)\sinh(h_1,k) + \cos(hk)\cosh(h_1,k), \end{aligned} \tag{7.3.3}$$

$\tilde{\varphi}(k)$ is the Fourier transform of the function $\varphi(\xi)$. Moreover, for each $z \in [0,H]$, $k \in (-\infty,\infty)$, $n = 0,1,2$, the inequalities

$$\left|\frac{\mathrm{d}^nU}{\mathrm{d}z^n}\right| \le 2\max\left(1, \frac{1}{\varkappa}\right)|k|^n|\tilde{\varphi}(k)| \tag{7.3.4}$$

hold which are the estimates of Lemma 7.2.2 in the (L–B) case.

Proof. We shall show that formulas (7.3.1)–(7.3.3) are obtained from formulas (7.2.2)–(7.2.15) in this case. For this purpose, we note that in our case

$$x = z, \qquad x_1 = z - h, \qquad \overset{+}{V}(z) \equiv \bar{V}(z) \equiv 1,$$

$$\overset{+}{g}(x_1) \equiv 0, \qquad \bar{g}(x) \equiv 0, \qquad \overset{+}{\alpha}_0 = \overset{+}{\alpha} = \bar{\alpha}_0 = \bar{\alpha} = 0,$$

$$\beta = \varkappa, \qquad \gamma = \delta_1 = \delta_2 = 1, \qquad x_h = h, \qquad x_H = H - h = h_1,$$

$$\varphi_e(x_1,s) = \cos(x_1s), \qquad \psi_e(x_1,s) = \sin(x_1s)/s,$$

so that

$$\Phi_e(s) = \varphi_e'(x_H, s) = -s \sinh_1 s, \qquad \Psi_e(s) = \psi_e'(x_H, s) = \cosh_1 s.$$

Since $\sin(ikx) = i \sinh(kx)$, $\cos(ikx) = \cosh(kx)$, then

$$\Phi_e(ik) = -ik \sin(ikh_1) = k \sinh(h_1 k),$$
$$\Psi_e(ik) = \cos(ikh_1) = \cosh(h_1 k).$$

Further, we have

$$\varphi_h(x, k) = \cos(kx), \qquad \psi_h(x, k) = \sin(kx)/k,$$

so, we obtain

$$\Phi_h(k) = \varphi_h'(h, k) = -k \sin(hk), \qquad \Psi_h(k) = \psi_h'(h, k) = \cos(hk).$$

Substituting these expression in formulas (7.2.2)–(7.2.5), we obtain

$$\begin{aligned}
A_1(k) &= \Phi_e(ik)\varphi_h(x_h, k) + \Psi_e(ik)\Phi_h(k) \\
&= k\{\cos(hk)\sinh(h_1 k) - \sin(hk)\cosh(h_1 k)\}, \\
A_2(k) &= \Phi_e(ik)\psi_h(x_h, k) + \Psi_e(ik)\Psi_h(k) \\
&= \sin(hk)\sinh(h_1 k) + \cos(hk)\cosh(h_1 k), \\
A_3(z, k) &= \Phi_e(ik)\psi_e(x_1, ik) - \Psi_e(ik)\varphi_e(x_1, ik) \\
&= k \sinh(h_1 k)\sin(ikx_1)/(ik) - \cosh(h_1 k)\cos(ikx_1) \\
&- \sinh(h_1 k)\sinh(kx_1) - \cosh(h_1 k)\cosh(kx_1) \\
&= -\cosh(k(x_1 - h_1)) = -\cosh(k(H - z)), \\
A_4(z, k) &\equiv A_1(k)\psi_h(x, k) - A_2(k)\varphi_h(x, h) \\
&= A_1(k)\sin(kz)/k - A_2(k)\cos(kz) \\
&= \sinh(h_1 k)\sin(k(z - h)) - \cosh(h_1 k)\cos(k(z - h)).
\end{aligned}$$

Hence, we obtain

$$W_e(x_1, k) = U_e(z, k) = \tilde{\varphi}(k)\,\frac{ikA_3(z, k)}{A_1(k) + i\varkappa k A_2(k)},$$
$$W_h(x, k) = U_h(z, k) = \tilde{\varphi}(k)\,\frac{ikA_4(z, k)}{A_1(k) + i\varkappa k A_2(k)},$$

i. e. formulas (7.3.1)–(7.3.3).

We show now, independently of the proof of Theorem 7.2.1, that the function $A(k) = A_1(k) + i\varkappa k A_2(k)$, for each $\varkappa \neq 0$, does not vanish for $-\infty < k < \infty$, $k \neq 0$, and the estimate

$$|A(k)|^2 \geq \min(1, \varkappa^2)k^2 \cosh(2h_1 k) \geq \min(1, \varkappa^2)k^2 \cosh^2(h_1 k)$$

holds for each $k \in (-\infty, \infty)$. This denotes that for $-\infty < k < \infty$, we have

$$A(k) \geq \min(1, |\varkappa|)|k| \cosh(h_1 k) \geq \min(1, |\varkappa|)|k| e^{h_1|k|}/2. \tag{7.3.5}$$

Really,

$$\begin{aligned}
|A(k)|^2 &= A_1^2(k) + \varkappa^2 k^2 A_2^2(k) \\
&\geq A_1^2 + k^2 A_2^2 = k^2\{\sinh^2(h_1 k) + \cosh^2(h_1 k)\} \\
&= k^2 \cosh(2h_1 k) \geq k^2 \cosh^2(h_1 k) \qquad \text{if} \quad |\varkappa| \geq 1, \\
|A(k)|^2 &\geq \varkappa^2\{A_1^2 + k^2 A_2^2\} = \varkappa^2 k^2 \cosh(2h_1 k) \\
&\geq \varkappa^2 k^2 \cosh^2(h_1 k), \qquad \text{if} \quad |\varkappa| \leq 1.
\end{aligned}$$

Hence, inequality (7.3.5) follows.

Since

$$|\cosh(k(H - z))/e^{kh_1/2}| \leq 2$$

for $h \leq z \leq H$ and taking into account (7.3.2) and (7.3.5) we obtain

$$\begin{aligned}
&|U_e(z, k)| = |\tilde{\varphi}(k)|\,|k \cosh(k(H - z))/A(k)| \\
&\leq |\tilde{\varphi}(k)|\,|\cosh(k(H - z))/e^{kh_1/2}|/\min(1, |\varkappa|) \leq 2\max(1, 1/|\varkappa|)\,|\tilde{\varphi}(k)|.
\end{aligned}$$

This is the estimate of type (7.3.4) for $h \leq z \leq H$, $n = 0$. Analogously, since

$$|A_4(z, k)| \leq |\sinh(h_1 k)| + |\cosh(h_1 k)| = e^{|k|h_1}$$

for $0 \leq z \leq h$, we obtain

$$|U_h(z, k)| = |\tilde{\varphi}(k)|\,|kA_4(z, k)/A(k)| \leq 2\max(1, 1/|\varkappa|)\,|\tilde{\varphi}(k)|.$$

This is the estimate of type (7.3.4) for $0 \leq z \leq h$, $n = 0$. The estimates of type (7.3.4) for $n = 1, 2$ are obtained similarly. Evidently estimates of type (7.3.4) obtained above independently of the proof of Theorem 7.2.1 coincide with the estimates of Lemma 7.2.2 for the functions W_e, W_h in the (L-B) case.

From formulas (7.3.3), estimates (7.3.5) follows immediately that the function $A(k)$ has the only zero (of the first order) in the real axis $-\infty <$

$k < \infty$, for $\varkappa \neq 0$. This zero is the point $k = 0$. This fact is established in § 7.2 in the general case.

As for the stability estimates (7.2.36) for the (L–B) case, we can set in them

$$C = 2\max(1, 1/|\varkappa|)$$

due to inequalities (7.3.4). □

7.4. INVERSE PROBLEMS. THE CASE $K(h+0) \neq 0$, $K(h-0) \neq 0$

7.4.1. Inverse problem with the data given in the boundary $z = H$.

Inverse problem 7.1. Suppose all the conditions of Theorem 7.2.1 hold and let for $-\infty < \xi < \infty$ the function $f(\xi)$ from condition (7.1.4) be given. Suppose also that we know the values $u(H,\xi)$ of solution $u(z,\xi)$ of the direct problem 7.1 with the numbers $H, \varkappa$ unknown. The function $K(z)(\overset{+}{V}(z))$ is unknown also in the segment $[h, H]$. The number h and the function $K(z)(\bar{V}(z))$ in the interval $[0, h]$ are known. We need to find the numbers $H, \varkappa$ and the function $K(z)$ in the segment $h \le z \le H$.

Solution of the inverse problem 7.1. Let $q(x)$ be a real function, continuous in a finite segment of definition $0 \le x \le l$, and

$$L(y) = -y'' + q(x)y$$

be the differential operator correspondent to this function. Denote by

$$S\{q(x), l, \varphi, \theta\}, \qquad (-\infty < \varphi, \theta < \infty)$$

the set of eigenvalues (the spectrum) of the classical (regular) Sturm—Liouville boundary-value problem

$$L(y) = \lambda y,$$
$$\cos\varphi y(0) + \sin\varphi y'(0) = 0, \qquad \cos\theta y(l) + \sin\theta y'(l) = 0. \tag{7.4.1}$$

It is known that the number l is determined by setting a certain spectrum $S\{q(x), l, \varphi, \theta\} = \{\lambda_n\}_1^\infty$, since (see Krein, 1961; Marchenko, 1952)

$$l^2 = \pi^2 \lim_{n\to\infty} \frac{n^2}{\lambda_n}.$$

Following the Marchenko theorem (Marchenko, 1952; Gasymov and Levitan 1964), the spectra $S\{q(x), l, \varphi_1, \theta\}$ and $S\{q(x), l, \varphi_2, \theta\}$ ($\operatorname{ctg}\varphi_1 \neq \operatorname{ctg}\varphi_2$) of two boundary-value problems with the same boundary conditions in the point $x = l$ and various boundary conditions in the point $x = 0$ define uniquely both the function $q(x)$ and the numbers $\operatorname{ctg}\varphi_1$, $\operatorname{ctg}\varphi_2$, and $\operatorname{ctg}\theta$.

It occurs that the data of the inverse problem 7.1 allow us to find the spectra of two boundary-value problems of type (7.4.1) with the same conditions in the point $x_1 = x_H$ and with different conditions in the point $x_1 = 0$:

$$-y'' + \overset{+}{q}(x_1)y = \lambda y, \qquad 0 \leq x_1 \leq x_H,$$
$$\cos\varphi y(0) + \sin\varphi y'(0) = 0, \qquad \cos\theta y(x_H) + \sin\theta y'(x_H) = 0; \tag{I}$$

$$y'' + \overset{+}{q}(x_1)y = \lambda y, \qquad 0 \leq x_1 \leq x_H,$$
$$y(0) = 0, \qquad \cos\theta y(x_H) + \sin\theta y'(x_H) = 0,, \tag{II}$$

where $\operatorname{ctg}\varphi = -\overset{+}{\alpha}_0$, $\operatorname{ctg}\theta = -\overset{+}{\alpha}$; the function $\overset{+}{q}(x_1)$ and the numbers x_H, $\overset{+}{\alpha}_0$, $\overset{+}{\alpha}$ are taken from system (7.2.16)–(7.2.21).

Really, by Theorem 7.2.1 and identity (7.2.22) we have

$$U(H,k) = -\overset{+}{v}{}^{1/2}(x_H)\,\frac{\gamma ik\tilde{\varphi}(k)}{A_1(k) + ik\beta A_2(k)}.$$

Inverting integral (7.2.1) for $z = H$ and taking into account the function $\varphi(\xi)$ is given, we define the function

$$cA(k) = cA_1(k) + ikc\beta A_2(k), \quad c = \{\gamma \overset{+}{v}{}^{1/2}(x_H)\}^{-1}$$

in an arbitrary segment $[a,b]$ of real axis $-\infty < k < \infty$.

Since the numbers c, β (until being unknown) and the functions $A_1(k)$, $A_2(k)$, $-\infty < k < \infty$ are real and $\beta \neq 0$, then, separating real and imaginary parts of the function $cA(k)$, we find the functions $cA_1(k)$ and $c\beta A_2(k)$ in the segment $[a,b]$. Define the number $\varkappa$ by the formula

$$\varkappa = c\beta A_2(0), \qquad (A_2(0) = \bar{v}^{-1/2}(0)\,\overset{+}{v}{}^{1/2}(x_H)),$$

which follows from equalities (7.2.25)–(7.2.27) and derive then the number $\beta = \varkappa\bar{v}(0)$ and the function $cA_2(k)$ in the segment $[a,b]$[iv].

[iv]If the point $k = 0\overline{\in}[a,b]$, then the value $c\beta A_2(0)$ may be defined if we extend analytically the entire function $c\beta A_2(k)$ from the segment $[a,b]$ into a certain neighborhood of the point $k = 0$.

Since the number h and the function $K(z)$ for $z \in [0, h]$ are known, we can determine the numbers $x_h, \bar{\alpha}_0, \bar{\alpha}$, the function $\bar{q}(x)$ and construct the functions $\varphi_h(x_h, k)$, $\psi_h(x_h, k)$, $\Phi_h(k)$, $\Psi_h(k)$ for $-\infty < k < \infty$.

Therefore, taking into account the identities

$$
\begin{aligned}
c\{A_1(k)\psi_h(x_h, k) - A_2(k)\varphi_h(x_h, k)\} &= -c\delta_2\Psi_e(s), \\
c\{A_1(k)\Psi_h(k) - A_2(k)\Phi_h(k)\} &= c\delta_1\Phi_e(s),
\end{aligned}
\tag{7.4.2}
$$

where $s = ik$, we can find the functions $c\delta_2\Psi_e(s)$ and $c\delta_1\Phi_e(s)$ in a certain segment $[ia, ib]$ of the imaginary axis of the complex plane s. Extending analytically (for example, by means of the Taylor series) the entire functions $c\delta_2\Psi_e(s)$ and $c\delta_1\Phi_e(s)$ from this segment into the real axis $-\infty < s < \infty$, we find them for $-\infty < s < \infty$. Therefore, all the real zeros of the functions $\Phi_e(s)$ and $\Psi_e(s)$ will be obtained. As it follows from the properties of self-adjoint Sturm—Liouville operator (Gasymov and Levitan, 1964; Levitan and Sargsyan, 1970, p. 13) the functions $\Phi_e(s)$ and $\Psi_e(s)$ have no more zeros; therefore, the real zeros consist the set of all zeros of these functions.

Note that the squares $\lambda = s^2$ of all zeros of the functions $\Phi_e(s)$ and $\Psi_e(s)$ form the spectra $S\{\overset{+}{q}(x_1), x_H, \varphi, \theta\}$ and $S\{\overset{+}{q}(x_1), x_H, 0, \theta\}$ of boundary-value problems (I) and (II) respectively. This follows from the theory of classical Sturm—Liouville problem (Gasymov and Levitan, 1964; Levitan and Sargsyan, 1970, p. 18). Following this theory, given these spectra, we find uniquely the numbers $x_H, \overset{+}{\alpha}_0, \overset{+}{\alpha}$ and the function $\overset{+}{q}(x_1)$ for $0 \le x_1 \le x_H$. To this end, we apply the known methods (Krein, 1951; Marchenko, 1951; Gasymov and Levitan, 1964). By the recovered values x_H, $\overset{+}{\alpha}_0$, $\overset{+}{\alpha}$, $\overset{+}{q}(x_1)$, construct the functions $\Phi_e(s)$, $\Psi_e(s)$. Then, taking into account formulas (7.4.2) we find the number $\delta_2/\delta_1 = \delta_2^2$, and then, we find the value $\overset{+}{v}(0) = \delta_2^2\bar{v}(x_h)$. From formula (7.2.27), for $\varphi_e(x_1, 0)$, we find the function $\overset{+}{v}(x_1)$ for $0 \le x_1 \le x_H$; and from the relations

$$
z = \int_0^{x_1} \overset{+}{v}(x_1)\, dx_1 + h, \quad \overset{+}{V}(z) = \overset{+}{v}(x_1) \quad \text{for} \quad z = z(x_1),\ H = z(x_H)
$$

we derive the number H and the function $K(z)$ for $h \le z \le H$.

7.4.2. Inverse problems with the data given in the boundary $z = h$.

Inverse problem 7.2. Let all the conditions of Theorem 7.2.1 hold and the function $f(\xi)$ be given for $-\infty < \xi < \infty$. Suppose also that one of the functions $u(h,\xi)$, $\partial u/\partial z|_{z=h}$ is set, where $u(z,\xi)$ solves direct problem 7.1 with unknown numbers $H, \varkappa$ and function $K(z)$ unknown in the segment $[h, H]$. The number h and the function $K(z)$ $(\bar{V}(z))$ in the interval $[0, h]$ are known also. We need to find these numbers $H, \varkappa$ and the function $K(z)$ for $h \leq z \leq H$.

Solution of inverse problem 7.2. Consider the case, when the function $u(h,\xi)$ is given (and the derivative $\partial u/\partial z|_{z=h}$ is unknown). Taking into account Theorem 7.2.1, we obtain

$$U(h,k) = -\overset{+}{v}{}^{1/2}(0)\,\frac{\gamma ik\tilde{\varphi}(k)\Psi_e(ik)}{A_1(k)+ik\beta A_2(k)}.$$

Inverting integral (7.2.1) for $z = h$, we find, first, the function

$$c_1(A_1(k)+ik\beta A_2(k))/\Psi_e(ik) \qquad (c_1 = \{\gamma\overset{+}{v}{}^{1/2}(0)\}^{-1})$$

in a certain interval of real axis $-\infty < k < \infty$. Then, separating the real and imaginary parts, we obtain the function

$$c_1A_1(k)/\Psi_e(ik) = c_1\{\delta_1\varphi_h(x_h,k)\Phi_e(ik)/\Psi_e(ik) + \delta_2\Phi_h(k)\}$$

and the number

$$\varkappa = \{[c_1\beta A_2(k)]/\Psi_e(ik)\}|_{k=0}.$$

Since the number h and the function $K(z)$ for $z \in [0,h]$ are known, we can construct the functions $\varphi_h(x_h,k)$ and $\Phi_h(k)$. Besides, the number $c_1\delta_2 = \{v(0)\bar{v}(x_h)\}^{-1/2}$ is known; therefore, the function

$$c_1\delta_1 m(s) = c_1\delta_1\Phi_e(s)/\Psi_e(s) \tag{7.4.3}$$

can be derived also in a certain segment of imaginary axis of the plane s. Using the method of analytical extension, define all the zeros of the functions $m(s)$ and $\{m(s)\}^{-1}$. The squares of these zeros form the spectra of problems (I) and (II) respectively. Then we follows the method of solving the inverse problem 7.1; but the value $\overset{+}{v}(0)$ we find by the formula

$$\overset{+}{v}(0) = \{c_1\delta_1\bar{v}^{1/2}(0)\bar{v}^{-1/2}(x_h)\}^{-1}$$

(deriving the numbers $c_1\delta_1$ from equality (7.4.3) beforehand).

The case, when the derivative $\partial u/\partial z|_{z=h}$ is given, is considered analogously. The distinction is that the number $\varkappa$ we derive from the formula

$$\varkappa = \tilde{c}_1\beta A_2(0)\,\{\overset{+}{v}(0)\,\overset{+}{v}(x_H)\}^{-1/2},$$

where $\tilde{c}_1 = \gamma^{-1}\overset{+}{v}{}^{1/2}(0)$, after recovering the function $\overset{+}{v}(x_1)$.

Inverse problem 7.3. Let all the conditions of Theorem 7.2.1 hold and the values $u(h,\xi)$ and $\partial u/\partial z|_{z=h}$ be given, where $u(z,\xi)$ is the solution of the direct problem 7.1. The numbers H, $\varkappa$ are unknown; $K(z)$ is unknown in each segment $[0,h]$ and $[h,H]$; the function $f(\xi)$ is unknown for $-\infty < \xi < \infty$. The number h is given. We need to find the number H and the function $K(z)$ in the segment $[h,H]$.

If in the inverse problem 7.3 the number h is unknown also, then, instead of H and $K(z)$ in the segment $[h,H]$ we must find the number $H-h$ and the function $\tilde{K}(z_1)$ for $0 \le z_1 \le H-h$. This function $\tilde{K}(z_1)$ is derived for $0 \le z_1 \le H-h$ by the formulas

$$z_1 = z - h, \qquad \tilde{K}(z_1) = K(z), \qquad z \in [h,H].$$

Suppose, in the inverse problem 7.3, the numbers $h, \varkappa$ are given and the function $K(z)$ is given in the segment $[0,h]$. Then, in addition, we must determine the function $f(\xi)$ for $-\infty < \xi < \infty$. Conversely, if the number h, the function $K(z)$ for $z \in [0,h]$, and the function $f(\xi)$ for $\xi \in (-\infty,\infty)$ are given, then, in addition, we need to find the number $\varkappa$.

The solution of this problem is based on determination of the spectra of the problems (I) and (II) also.

7.4.3. Inverse problems with the data in the boundary $z = 0$.

Inverse problem 7.4. Let all the conditions of Theorem 7.2.1 hold. Suppose that for $-\infty < z < \infty$ are given: the function $f(\xi)$; the values $u(0,\xi)$ of the solution $u(z,\xi)$ of the direct problem 7.1. The numbers H, $\varkappa$ and the function $K(z)$ in the segment $[h,H]$ are unknown. The number h and the function $K(z)$ in the segment $[0,h]$ are given. We need to find the numbers H, $\varkappa$ and the function $K(z)$ for $h \le z \le H$.

Solution of the inverse problem 7.4. By Theorem 7.2.1 we have

$$U(0,k) = -\bar{v}(0)\,\frac{ik\tilde{\varphi}(k)A_2(k)}{A_1(k) + ik\beta A_2(k)}.$$

Inverting integral (7.2.1) for $z = 0$, we find, at first, the function

$$\{A_1(k) + ik\beta A_2(k)\}/A_2(k),$$

and then, we find the function

$$R(k) = \frac{A_1(k)}{A_2(k)} = \frac{\delta_1 m(s)\varphi_h(x_h,k) + \delta_2\Phi_h(k)}{\delta_1 m(s)\psi_h(x_h,k) + \delta_2\Psi_h(k)}, \qquad m(s) = \frac{\Phi_e(s)}{\Psi_e(s)}$$

in a certain segment of the real axis of the complex plane k and the numbers β, $\varkappa = \beta/\bar{v}(0)$. Then we construct the functions $\varphi_h(x_h, k)$, $\psi_h(x_h, k)$, $\Phi_h(k)$, $\Psi_h(k)$ and derive the function $\delta_1^2 m(s)$ in a certain segment of the imaginary axis of the complex plane s from the formula

$$\frac{\delta_1}{\delta_2} m(s) = -\frac{\Phi_h(k) - R(k)\Psi_h(k)}{\varphi_h(x_h, k) - R(k)\psi_h(x_h, k)}.$$

Further, we follow the method of solution of the inverse problems 7.1 and 7.2.

Inverse problem 7.5. Let all the conditions of Theorem 7.2.1 hold and the values $u(0, \xi)$ and $\partial u/\partial z|_{z=0}$ be given for $-\infty < \xi < \infty$. The function $u(z, \xi)$ solves the direct problem 7.1. The numbers $H, \varkappa$, the function $K(z)$ for $[h, H]$, and the function $f(\xi)$ for $-\infty < \xi < \infty$ are unknown. The number h and the function $K(z)$ in the segment $[0, h]$ are given. We need to find the number H and the function $K(z)$ for $h \le z \le H$.

This problem we solve similarly to the inverse problem 7.4.

From the method of solution of inverse problems 7.1–7.5 follows the theorem.

Theorem 7.4.1. *All the inverse problems 7.1–7.5 have the unique solution.*

7.5. THE GENERAL CASE

In § 7.2–7.4 we have considered the particular case when the coefficient $K(z)$ in equation (7.0.1) does not vanished in the line $z = h$, where the equation changes the type. Therefore, the coefficient $K(z)$ has a finite jump. We shall consider here the general case of the behavior of the coefficient $K(z)$ passing through the point $z = h$ from the domain of ellipticity of (7.0.1) to the domain of hyperbolicity. In this case the coefficient $K(z)$ may change both continuously, vanishing in the point $z = h$ and discontinuously (in this case $K(z)$ has the jump). In the continuous case, the coefficient $K(z)$ may vanish in the point $z = h$ with arbitrary power which may be different from the both sides. In the case of the jump, $K(z)$ may vanish only from the one side of the point $z = h$, or not vanish at all. The general case includes the case when $K(h+0) \neq 0$; $K(h-0) \neq 0$ considered in § 7.2–7.4 as a particular case. The representation of solution of the direct problem 7.1 in the general case gives the following theorem.

Theorem 7.5.1. *Set in condition (7.1.4)*

$$f(\xi) = \frac{d\varphi}{d\xi} \qquad (\varphi(0) = 0);$$

let the number $\varkappa \neq 0$, $\varkappa \in (-\infty, \infty)$;

(A) the function $K(z)$ *may be represented as follows:*

$$\begin{aligned} |K(z)| &= (h-z)^{\nu_1}\tau_1^2(z), \qquad 0 \le z \le h, \\ K(z) &= (z-h)^{\nu_2}\tau_2^2(z), \qquad h \le z \le H, \end{aligned} \tag{7.5.1}$$

where ν_1 *and* ν_2 *are arbitrary nonnegative numbers and the functions* $\tau_1^2(z)$, $\tau_2^2(z)$ *are positive and have continuous second derivative in the intervals* $[0, h]$ *and* $[h, H]$ *respectively.*

Let, moreover,

(B) the function $\varphi(\xi)$ *and its derivatives up to the fourth order be continuous and absolutely integrable in the line* $-\infty < \xi < \infty$;

(C) the function $\xi\varphi(\xi)$ *be absolutely integrable in the line* $-\infty < \xi < \infty$.

Then the solution $u(z, \xi)$ *of the direct problem 7.1 exists and has the representation*

$$u(z, \xi) = \int_{-\infty}^{+\infty} U(z, k) e^{ik\xi}\, dk, \tag{7.5.2}$$

$$U(z, k) = \begin{cases} U_e(z_2, k), & z_2 = z - h,\ h \le z \le H,\ 0 \le z_2 \le h_1, \\ U_h(z, k), & 0 \le z \le h, \end{cases} \tag{7.5.3}$$

where the integration goes along the real axis of the complex plane k *and*

$$\begin{aligned} U_e(z_2, k) &= \frac{ik\Phi(k)\{\varphi_e'(h_1, ik)\psi_e(z_1, ik) - \psi_e'(h_1, ik)\varphi_e(z_1, ik)\}}{A_1(k) + ik\varkappa A_2(k)}, \\ U_h(z, k) &= \frac{ik\Phi(k)\{A_1(k)\psi_h(z, k) - A_2(k)\varphi_h(z, k)\}}{A_1(k) + ik\varkappa A_2(k)}, \end{aligned} \tag{7.5.4}$$

$$\begin{aligned} A_1(k) &= \varphi_h(h, k)\varphi_e'(h_1, ik) + \varphi_h'(h, k)\psi_e'(h_1, ik), \\ A_2(k) &= \psi_h(h, k)\varphi_e'(h_1, ik) + \psi_h'(h, k)\psi_e'(h_1, ik). \end{aligned} \tag{7.5.5}$$

Here $h_1 = H - h$; *the functions* $\varphi_e(z_2, s)$ *and* $\psi_e(z_2, s)$ *solve the equation*

$$y'' + s^2\tilde{K}(z_2)y = 0, \qquad 0 \le z_2 \le h_1, \tag{7.5.6}$$

where $\tilde{K}(z_2) = K(z)$ *for* $z_2 = z - h$, $h \le z \le H$, *with the initial conditions*

$$\varphi_e(0,s) = 1, \qquad \varphi_e'(0,s) = 0; \tag{7.5.7}$$
$$\psi_e(0,s) = 0, \qquad \psi_e'(0,s) = 1. \tag{7.5.8}$$

The functions $\varphi_h(z,k)$ *and* $\psi_h(z,k)$ *solve the equation*

$$y'' + k^2|K(z)|y = 0, \qquad 0 \le z \le h \tag{7.5.9}$$

with the initial condition

$$\varphi_h(0,k) = 1, \qquad \varphi_h'(0,k) = 0; \tag{7.5.10}$$
$$\psi_h(0,k) = 0, \qquad \psi_h'(0,k) = 1. \tag{7.5.11}$$

The function $\Phi(k)$ *is the Fourier transform of the function* $\varphi(\xi)$; i *is the imaginary unity. The prime denotes the differentiation either with respect to* z *or* z_2. *The integrals in equality (7.5.2) and in the equality*

$$\frac{\partial u}{\partial z} = \int_{-\infty}^{+\infty} \frac{\mathrm{d}U}{\mathrm{d}z} e^{ik\xi} \, \mathrm{d}k$$

are Fouriers integrals (i. e., the inversion formulas hold).

Thus the formulas of Theorem 7.5.1 give the explicit expression of the solution $u(z,\xi)$ of the direct problem 7.1 via the fundamental systems of solutions of two string equations with a nonnegative density. The coefficient of the first equation from (7.5.6) and the system of solutions $\varphi_e(z_1,s)$, $\psi_e(z_1,s)$ are constructing by the values of the function $K(z)$ in the domain of ellipticity of equation (7.0.1). The coefficient of the second equation from (7.5.9) and the system of solutions $\varphi_h(z,k)$, $\psi_h(z,k)$ is constructing by the values of $K(z)$ in the domain where the equation (7.0.1) is hyperbolic.

Note that if the numbers ν_1, ν_2 are not equal to zero, then the function $K(z)$ is continuous in the point $z = h$ and is equal to zero in this point. The other variants of behaviour of the coefficient $K(z)$ in the point $z = h$ may be as follows:

1) $\nu_1 = 0$, $\nu_2 \ne 0$ and, therefore, $K(h+0) = 0$, $K(h-0) \ne 0$;
2) $\nu_1 \ne 0$, $\nu_2 = 0$; therefore $K(h+0) \ne 0$, $K(h-0) = 0$;
3) $\nu_1 = \nu_2 = 0$; therefore $K(h+0) \ne 0$, $K(h-0) \ne 0$. This case is considered in § 7.2–7.4.

In all three cases 1)–3) the function $K(z)$ has the jump in the point $z = h$. Thus, Theorem 7.5.1 envelops all four variants.

Proof of Theorem 7.5.1. Note that the functions $U_e(z_1,k)$, $U_h(z,k)$ defined by formulas (7.5.4)–(7.5.11) satisfy the following equations and conditions

$$\frac{\mathrm{d}^2U_h}{\mathrm{d}z^2}+k^2|K(z)|U_h=0, \qquad 0\le z\le h, \tag{7.5.12}$$

$$\frac{\mathrm{d}^2U_e}{\mathrm{d}z_2^2}+s^2\tilde{K}(z_2)U_e=0, \qquad 0\le z_2\le h_1, \quad s=ik, \tag{7.5.13}$$

$$\frac{\mathrm{d}U_e}{\mathrm{d}z_2}\Big|_{z_2=h_1}=0, \tag{7.5.14}$$

$$U_e(0,k)=U_h(h,k), \tag{7.5.15}$$

$$\frac{\mathrm{d}U_e}{\mathrm{d}z_2}\Big|_{z_2=0}=\frac{\mathrm{d}U_h}{\mathrm{d}z}\Big|_{z=h}, \tag{7.5.16}$$

$$\left\{\frac{\mathrm{d}U_h}{\mathrm{d}z}-ik\varkappa U_h\right\}\Big|_{z=0}=ik\Phi(k). \tag{7.5.17}$$

Equations (7.5.12), (7.5.13) and conditions (7.5.14)–(7.5.17) are set in the real axis $-\infty<k<\infty$ of the complex plane k; therefore, equation (7.5.13) is set in the imaginary axis of the complex plane $s=ik$.

Really, taking into account the condition (A), we see that the functions $|K(z)|$ and $\tilde{K}(z_2)$ are continuous and nonnegative in the segments $[0,h]$ and $[0,h_1]$ respectively. Therefore, equation (7.5.12) has the solutions $\varphi_h(z,k)$, $\psi_h(z,k)$ respectively for initial conditions (7.5.10), (7.5.11). The equation (7.5.13) has the solutions $\varphi_e(z_2,s)$, $\psi_e(z_2,s)$ respectively for initial conditions (7.5.7), (7.5.8) with the following properties (see Levitan and Sargsyan, 1970; Kats and Krein, 1964; Alekseev and Megrabov, 1974). For each fixed $z\in[0,h]$, the functions $\varphi_h(z,k),\varphi_h'(z,k),\psi_h(z,k),\psi_h'(z,k)$ are entire real functions dependent on k (this means that are real for real k). These functions are continuous with respect to (z,k) for $z\in[0,h]$ and for each complex k. The functions $\varphi_h(z,k),\psi_h(z,k)$ form the fundamental system of solutions of equation (7.5.12). Their Wronskian

$$\psi_h'(z,k)\varphi_h(z,k)-\psi_h(z,k)\varphi_h'(z,k)\equiv 1. \tag{7.5.18}$$

For negative $\lambda=k^2$, the functions $\varphi_h(z,k)$, $\psi_h(z,k)$, $\varphi_h'(z,k)$, $\psi_h'(z,k)$ attain nonnegative values everywhere in the segment $[0,h]$ and do not decrease in $[0,h]$.

The functions $\varphi_e(z_2,s)$, $\psi_e(z_2,s)$ have the analogous properties.

Represent the solution $\{U_e(z_1,k), U_h(z,k)\}$ in the form

$$U_e(z_2,k) = C_1(k)\varphi_e(z_2,k) + C_2(k)\psi_e(z_2,k),$$
$$U_h(z,k) = D_1(k)\varphi_h(z,k) + D_2(k)\psi_h(z,k)$$

and define the functions $C_1(k)$, $C_2(k)$, $D_1(k)$, $D_2(k)$from the conditions (7.5.14)–(7.5.17). As a result, we obtain formulas (7.5.4), (7.5.5) for the functions U_e, U_h.

The following lemma holds.

Lemma 7.5.1. *The functions $A_1(k)$, $A_2(k)$ defined by formulas (7.5.5) do not vanish simultaneously for each $k \in (-\infty, \infty)$.*

Proof. Suppose that for a certain k we have

$$A_1(k) = A_2(k) = 0.$$

Then the functions $A_1(k)\psi_h(h,k)$ and $A_2(k)\varphi_h(h,k)$ also are equal to zero. Therefore, for such k, we have

$$A_1(k)\psi_h(h,k) - A_2(k)\varphi_h(h,k) = 0. \tag{7.5.19}$$

Taking into account identity (7.5.18), we obtain

$$A_1(k)\psi_h(h,k) - A_2(k)\varphi_h(h,k) \equiv -\psi_e'(h_1, ik) = 0.$$

Therefore, equality (7.5.19) is impossible, if the number $s = ik$ is not a zero of the function $\psi_e'(h_1, s)$.

Let $s = ik$ be the zero of the function $\psi_e'(h_1, s)$. Then, in the point k, we have

$$A_1(k) = \varphi_h(h,k)\varphi_e'(h_1, ik), \qquad A_2(k) = \psi_h(h,k)\varphi_e'(h_1, ik).$$

Assume that $A_1(k) = A_2(k) = 0$. Then

$$A_1(k)\psi_h'(h,k) - A_2(k)\varphi_h'(h,k) = 0.$$

Taking into account identity (7.5.18), we obtain

$$A_1(k)\psi_h'(h,k) - A_2(k)\varphi_h'(h,k) \equiv \varphi_e'(h_1, ik) = 0.$$

So, $\varphi_e'(h_1, ik) = 0$. Therefore, in the point k, the two equalities hold $\varphi'(h_1, s) = 0$ and $\psi'(h_1, s) = 0$. However, this contradicts the identity

$$\psi_e'\varphi_e - \varphi_e'\psi_e \equiv 1.$$

The lemma is proved. □

Since the number $\varkappa$ and the functions $A_1(k)$, $A_2(k)$ are real for $-\infty < k < \infty$, then Lemma 7.5.1 yields the following corollary.

Corollary 7.5.1. *The function*

$$A(k) = A_1(k) + ik\varkappa A_2(k)$$

does not vanish for each $\varkappa \neq 0$ and $k \neq 0$, $k \in (-\infty, \infty)$.

The explicit formulas

$$\varphi_e(z_2, 0) = 1, \quad \psi_e(z_2, 0) = z_2, \quad \varphi_h(z, 0) = 1, \quad \psi_h(z, 0) = z,$$

yield

$$A_1(0) = 0, \qquad A_2(0) = 1.$$

So, the point $k = 0$ is the zero of the first order of the function $A(k)$. Therefore, the functions $U_e(z_2, k)$, $U_h(z, k)$ are continuous for $-\infty < k < \infty$.

Obtain now the estimates for behavior the functions $U_e(z_2, k)$ and $U_h(z, k)$ (and for their derivatives with respect to z_2 and z respectively) for $|k| \to \infty$, $k \in (-\infty, \infty)$. Taking into account the relations

$$\begin{aligned} U_e^{(n)}(z_2, -k) &= \bar{U}_e^{(n)}(z_2, k), \\ U_h^{(n)}(z, -k) &= \bar{U}_h^{(n)}(z, k) \qquad (n = 0, 1, 2), \end{aligned} \tag{7.5.20}$$

it suffices to obtain the estimates for the real semiaxis $k > 0$ (imaginary semiaxis $\operatorname{Im} s > 0$ of the complex parameter $s = ik$).

For this purpose, we apply the asymptotic formulas obtained by Langer (1931) for certain particular solutions of the ordinary linear differential equation

$$u''(x) + \{\rho^2 w^2(x) - \chi(x)\} u(x) = 0, \quad x \in J, \tag{7.5.21}$$

where ρ is a complex parameter, J is a finite segment $0 \leq x \leq x_c < \infty$, and the function $w^2(x)$ has $x = 0$ as a zero. The prime in (7.5.21) denotes the differentiation with respect to x.

Formulas of R. E. Langer.

Suppose that the following conditions hold relatively the coefficients $w^2(x)$ and $\chi(x)$ of equation (7.5.21):

(1) the function $w^2(x)$ vanishes in the point $x = 0$ with the degree ν, where ν is any nonnegative number;

(2) the function $w^2(x)$ has no more zeros in the segment J except the point $x = 0$;

(3) the function $x^{-\nu}w^2(x)$ is real positive and has the second continuous derivative with respect to x in the segment J;

(4) the function $\chi(x)$ is real and continuous in the segment J.

The Taylor expansion yields the formula

$$w^2(x) = x^\nu(\alpha_0 + \alpha_1 x + \alpha_2(x)x^2), \qquad \alpha_0 \neq 0 \tag{7.5.22}$$

where the coefficients $\alpha_0, \alpha_1, \alpha_2$ are real, α_0 and α_1 are the constants and $\alpha_2(x)$ is a continuous function in J.

We define the functions $\Phi(x)$ and $\Psi(x)$ as follows

$$\Phi(x) = \int_0^x w(t)\,dt, \qquad \Psi(x) = \frac{\{\Phi(x)\}^{1/2-\mu}}{\{w(x)\}^{1/2}},$$

where the constant μ defines by the formula $\mu = 1/(\nu+2)$, and set $\xi = \rho\Phi(x)$. Taking into account the conditions imposed on the function $w^2(x)$, we see that each of the functions $x^{-1/(2\mu)}\Phi(x)$, $\Psi(x)$ is real positive and has the continuous second derivative in the segment J. Therefore, applying the Taylor theorem, we obtain the following formulas

$$\begin{aligned} \Phi(x) &= x^{1/(2\mu)}\{\beta_0 + \beta_1 x + \beta_2(x)x^2\}, \qquad \beta_0 \neq 0, \\ \Psi(x) &= \{\delta_0 + \delta_1 x + \delta_2(x)x^2\}, \qquad \delta_0 \neq 0. \end{aligned} \tag{7.5.23}$$

The coefficients $\beta_2(x), \delta_2(x)$ are continuous functions and $\beta_0, \beta_1, \delta_0, \delta_1$ are the constants which can be derived easily from equality (7.5.22) via the constants α_0, α_1.

It should be noted that for each $\nu \geq 0$, the constant μ satisfies the inequality $\mu \leq 1/2$.

If the function $\omega^2(x)$ does not vanish in the segment J, this corresponds to the case $\nu = 0$, $\mu = 1/2$.

In the paper of Langer (1931) the following estimates are obtained for certain particular solutions $u_1(x,\rho)$, $u_2(x,\rho)$ of equation (7.5.21), which form the fundamental system.

Suppose $u_1(x,\rho)$ and $u_2(x,\rho)$ solve the equation (7.5.21) for the following initial conditions

$$\begin{aligned} u_1(0,\rho) &= 2^\mu\Psi(0)/\Gamma(1-\mu), \qquad \bar{u}_1(0,\rho) = 0, \\ u_2(0,\rho) &= 0, \qquad \bar{u}_2(0,\rho) = 2^{1-\mu}\Psi(0)/(i\Gamma(\mu)), \end{aligned} \tag{7.5.24}$$

where

$$\bar{u}_j(x,\rho) = \Big\{u'_j(x,\rho) - \frac{\Psi'(x)}{\Psi(x)}\,u_j(x,\rho)\Big\}\frac{\Psi^2(x)}{i\rho^{2\mu}}, \qquad j=1,2,$$

and Γ is the gamma-function.

The prime here (as in equation (7.5.21)) denotes the differentiation with respect to x.

Then there exist such numbers $N>0$ and $\rho_N>0$, that for $|\xi|\le N$, $|\rho|>\rho_N$, $j=1,2$

$$\begin{aligned} u_j(x) &= \Psi(x)\xi^{\mu}J_{\mp\mu}(\xi) + O(|\rho|^{-4\mu}),\\ \bar{u}_j(x) &= \pm i\Psi(x)\xi^{1-\mu}J_{\pm(1-\mu)}(\xi) + O(|\rho|^{-4\mu}). \end{aligned} \tag{7.5.25}$$

Here J is the Bessel function (of the first kind).

For $|\xi|>N$, $\rho>\rho_N$, $j=1,2$, $x>0$, $-\pi/2<\arg\rho\le\pi/2$ we have

$$u_j(x,\rho) = \Psi(x)\xi^{\mu-1/2}\Big\{c_{j,1}e^{i\xi} + c_{j,2}e^{-i\xi} + O\Big(\frac{e^{|\operatorname{Im}\xi|}}{|\rho|^{\sigma_1}}\Big) + O\Big(\frac{e^{|\operatorname{Im}\xi|}}{|\xi|}\Big)\Big\}$$

$$\bar{u}_j(x,\rho) = \Psi(x)\xi^{1/2-\mu}\Big\{c_{j,1}e^{i\xi} - c_{j,2}e^{-i\xi} + O\Big(\frac{e^{|\operatorname{Im}\xi|}}{|\rho|^{\sigma_1}}\Big) + O\Big(\frac{e^{|\operatorname{Im}\xi|}}{|\xi|}\Big)\Big\},$$

where

$$c_{j,1} = e^{-1/2(1/2\mp\mu)\pi i}/(2\pi)^{1/2}, \quad c_{j,2} = e^{1/2(1/2\mp\mu)\pi i}/(2\pi)^{1/2} = \bar{c}_{j,1}. \tag{7.5.26}$$

Everywhere in formulas (7.5.25), (7.5.26) the upper sign is related to the value $j=1$ and the lower sign to the value $j=2$. The function $O(z)$ is a function dependent on the two variables x, ρ, satisfying the condition $|O(z)|<Cz$, where the constant C is independent on x, ρ. The constant σ_1 is positive, $\sigma_1\le 4\mu$.

All these formulas hold for each value $\nu\ge 0$, i.e., for each $\mu\in(0,1/2]$. The values $\nu=0$, $\mu=1/2$, as was noted, corresponds to the case when $w^2(x)$ has no zeros in the segment J.

Note that formulas (7.5.26) are obtained by R. E. Langer not only for the half-plane

$$\Sigma_0: \qquad -\pi/2<\arg\rho\le\pi/2,$$

but for each half-plane

$$\Sigma_n: \qquad (n-1/2)\pi<\arg\rho\le(n+1/2)\pi, \qquad n=0,\pm1,\pm2,\dots.$$

However, for our purpose we need to use only the formulas for u_1, u_2 in the half-plane Σ_0. Really, taking into account equalities (7.5.20), it suffices to obtain the estimates for the functions U_h, U_e in the real semiaxis $k > 0$ of the complex parameter k (in the imaginary semiaxis $\operatorname{Im} s > 0$ of the complex parameter $s = ik$). Hence, we need to use the Langer formula for u_1, u_2 only in the real semiaxis $\rho > 0$ ($\arg \rho = 0$) and in the imaginary semiaxis $\operatorname{Im} \rho > 0$ ($\arg \rho = \pi/2$). Both these semiaxes belong to the half-plane Σ_0.

7.5.1. Asymptotic formulas for the functions $\varphi(x, \rho)$ and $\psi(x, \rho)$. Suppose that $\varphi(x, \rho)$ and $\psi(x, \rho)$ solve equation (7.5.21) for the initial conditions

$$\varphi(0, \rho) = 1, \quad \varphi'(0, \rho) = 0, \quad \psi(0, \rho) = 0, \quad \psi'(0, \rho) = 1. \tag{7.5.27}$$

We express the functions $\varphi(x, \rho), \psi(x, \rho), \varphi'(x, \rho), \psi'(x, \rho)$ via the functions $u_1(x, \rho)$, $u_2(x, \rho)$:

$$\begin{aligned} \varphi(x, \rho) &= \frac{\Gamma(1-\mu)}{2^{\mu}\Psi(0)} u_1(x, \rho) - \frac{\Gamma(\mu)\Psi'(0)}{2^{1-\mu}\rho^{2\mu}} u_2(x, \rho), \\ \psi(x, \rho) &= \frac{\Gamma(\mu)\Psi(0)}{2^{1-\mu}\rho^{2\mu}} u_2(x, \rho), \end{aligned} \tag{7.5.28}$$

$$\begin{aligned} \varphi'(x, \rho) = {} & \frac{\Gamma(1-\mu)}{2^{\mu}\Psi(0)} \frac{i\rho^{2\mu}}{\Psi^2(x)} \bar{u}_1(x, \rho) - \frac{\Gamma(\mu)\Psi'(0)}{2^{1-\mu}} \frac{i}{\Psi^2(x)} \bar{u}_2(x, \rho) \\ & + \frac{\Gamma(1-\mu)}{2^{\mu}\Psi(0)} \frac{\Psi'(x)}{\Psi(x)} u_1(x, \rho) - \frac{\Gamma(\mu)\Psi'(0)}{2^{1-\mu}} \frac{1}{\rho^{2\mu}} \frac{\Psi'(x)}{\Psi(x)} u_2(x, \rho), \end{aligned}$$

$$\psi'(x, \rho) = \frac{\Gamma(\mu)\Psi(0)}{2^{1-\mu}} \frac{i}{\Psi^2(x)} \bar{u}_2(x, \rho) + \frac{\Gamma(\mu)\Psi(0)}{2^{1-\mu}} \frac{1}{\rho^{2\mu}} \frac{\Psi'(x)}{\Psi(x)} u_2(x, \rho).$$

Substituting the Langer formulas (7.5.25), (7.5.26) into (7.5.28), we obtain the asymptotic formulas for the functions $\varphi(x, \rho)$, $\psi(x, \rho)$, $\varphi'(x, \rho)$, $\psi'(x, \rho)$. To this end it suffices to use these formulas only for the real ($\arg \rho = 0$) and purely imaginary ($\arg \rho = \pi/2$) values of the complex parameter ρ.

For the real value ρ ($\rho > 0$), taking into account (7.5.26), for $|\xi| > N$, $|\rho| > \rho_N$, $x > 0$ we have

$$\begin{aligned} u_1(x, \rho) &= \sigma\Psi(x)\xi^{\mu-1/2}[\cos\{\xi - (1/2 - \mu)\pi/2\} + \ldots], \\ u_2(x, \rho) &= \sigma\Psi(x)\xi^{\mu-1/2}[\cos\{\xi - (1/2 + \mu)\pi/2\} + \ldots], \\ \bar{u}_1(x, \rho) &= \sigma\Psi(x)\xi^{1/2-\mu}[i \sin\{\xi - (1/2 - \mu)\pi/2\} + \ldots], \\ \bar{u}_2(x, \rho) &= \sigma\Psi(x)\xi^{1/2-\mu}[i \sin\{\xi - (1/2 + \mu)\pi/2\} + \ldots], \end{aligned}$$

where $\sigma = \{2/\pi\}^{1/2}$. The sums of the terms

$$O(1/|\rho|^{\sigma_1}), \qquad O(1/|\xi|) \tag{7.5.29}$$

we denote by dots.

Hence it follows that for real $\rho > 0$ for the functions $\varphi(x,\rho)$, $\psi(x,\rho)$, $\varphi'(x,\rho)$, $\psi'(x,\rho)$ the following formulas hold for $|\xi| > N$, $|\rho| > \rho_N$, $x > 0$:

$$\begin{aligned}
\varphi(x,\rho) &= \sigma\,\frac{\Gamma(1-\mu)}{2^{\mu}\Psi(0)}\,\Psi(x)\xi^{\mu-1/2}\Big\{\cos\Big\{\xi-\frac{\pi}{2}\Big(\frac{1}{2}-\mu\Big)\Big\}+\dots\Big\},\\
\psi(x,\rho) &= \sigma\,\frac{\Gamma(\mu)\,\Psi(0)}{2^{1-\mu}}\Psi(x)\xi^{\mu-1/2}\rho^{-2\mu}\Big\{\cos\Big\{\xi-\frac{\pi}{2}\Big(\frac{1}{2}+\mu\Big)\Big\}+\dots\Big\},\\
\varphi'(x,\rho) &= -\sigma\,\frac{\Gamma(1-\mu)}{2^{\mu}\Psi(0)}\,\frac{1}{\Psi(x)}\,\xi^{1/2-\mu}\rho^{2\mu}\Big\{\sin\Big\{\xi-\frac{\pi}{2}\Big(\frac{1}{2}-\mu\Big)\Big\}+\dots\Big\},\\
\psi'(x,\rho) &= -\sigma\,\frac{\Gamma(\mu)\Psi(0)}{2^{1-\mu}}\,\frac{1}{\Psi(x)}\,\xi^{1/2-\mu}\Big\{\sin\Big\{\xi-\frac{\pi}{2}\Big(\frac{1}{2}+\mu\Big)\Big\}+\dots\Big\}.
\end{aligned} \tag{7.5.30}$$

The sums of the terms

$$O(|\rho|^{-2\mu}),\ O(|\rho|^{-1}),\ O(|\rho|^{-\sigma_1}),\ O(|\xi|^{-1}),\ O(|\rho|^{-2\mu}|\xi|^{1-2\mu}) \tag{7.5.31}$$

are denoted here by dots.

If ρ is purely imaginary, $\operatorname{Im}\rho > 0$, $\rho = i|\rho|$, $\arg\rho = \pi/2$, taking into account the equalities

$$\xi = \Phi(x)|\rho|e^{i\pi/2} = |\xi|e^{i\pi/2},$$
$$\xi^{\mu-1/2} = |\xi|^{\mu-1/2}e^{(\mu-1/2)\pi i/2}, \qquad \xi^{1/2-\mu} = |\xi|^{1/2-\mu}e^{(1/2-\mu)\pi i/2}$$

we obtain from formulas (7.5.26)

$$\begin{aligned}
u_1(x,\rho) &= (\sigma/2)\Psi(x)|\xi|^{\mu-1/2}e^{|\xi|}\{1+\dots\},\\
u_2(x,\rho) &= (\sigma/2)\Psi(x)|\xi|^{\mu-1/2}e^{\mu\pi i}e^{|\xi|}\{1+\dots\},\\
\bar{u}_1(x,\rho) &= -(\sigma/2)\Psi(x)|\xi|^{1/2-\mu}e^{(1/2-\mu)\pi i}e^{|\xi|}\{1+\dots\},\\
\bar{u}_2(x,\rho) &= -(\sigma/2)\Psi(x)|\xi|^{1/2-\mu}ie^{|\xi|}\{1+\dots\}.
\end{aligned}$$

Here the sums of the terms of form (7.5.29) are denoted by dots, as before.

Hence it follows, that for ρ purely imaginary ($\rho = i|\rho|$), for $|\xi| > N$, $|\rho| > \rho_N$, $x > 0$ the formulas below hold

$$\begin{aligned}\varphi(x,\rho) &= \frac{\sigma}{2}\frac{\Gamma(1-\mu)}{2^{\mu}\Psi(0)}\Psi(x)|\xi|^{\mu-1/2}e^{|\xi|}\{1+\ldots\},\\ \psi(x,\rho) &= \frac{\sigma}{2}\frac{\Gamma(\mu)\Psi(0)}{2^{1-\mu}}\Psi(x)|\rho|^{-2\mu}|\xi|^{\mu-1/2}e^{|\xi|}\{1+\ldots\},\\ \varphi'(x,\rho) &= \frac{\sigma}{2}\frac{\Gamma(1-\mu)}{2^{\mu}\Psi(0)}\frac{|\rho|^{2\mu}}{\Psi(x)}|\xi|^{1/2-\mu}e^{|\xi|}\{1+\ldots\},\\ \psi'(x,\rho) &= \frac{\sigma}{2}\frac{\Gamma(\mu)\Psi(0)}{2^{1-\mu}}\frac{1}{\Psi(x)}|\xi|^{1/2-\mu}e^{|\xi|}\{1+\ldots\},\end{aligned} \tag{7.5.32}$$

where the sums of the terms of form (7.5.31) are denoted by dots.

The formulas (7.5.30), (7.5.32) obtained for the functions $\varphi(x,\rho)$, $\psi(x,\rho)$, $\varphi'(x,\rho)$, $\psi'(x,\rho)$ are related to the values $|\xi| > N$. We shall write now the asymptotic formulas for these functions for $|\xi| \le N$. For this purpose, we use the formulas (7.5.25) and (7.5.28).

As a result we obtain for $|\xi| \le N$, $|\rho| > \rho_N$, $\xi \in \Sigma_0(x > 0, -\pi/2 < \arg\rho \le \pi/2)$ the formulas:

$$\begin{aligned}\varphi(x,\rho) &= \frac{\Gamma(1-\mu)}{2^{\mu}\Psi(0)}\Psi(x)\xi^{\mu}J_{-\mu}(\xi)+\ldots,\\ \psi(x,\rho) &= \frac{\Gamma(\mu)\Psi(0)}{2^{1-\mu}}\Psi(x)\rho^{-2\mu}\xi^{\mu}J_{\mu}(\xi)+\ldots,\\ \varphi'(x,\rho) &= -\frac{\Gamma(1-\mu)}{2^{\mu}\Psi(0)}\frac{1}{\Psi(x)}\rho^{2\mu}\xi^{1-\mu}J_{1-\mu}(\xi)+\ldots,\\ \psi'(x,\rho) &= \frac{\Gamma(\mu)\Psi(0)}{2^{1-\mu}}\frac{1}{\Psi(x)}\xi^{1-\mu}J_{-(1-\mu)}(\xi)+\ldots,\end{aligned} \tag{7.5.33}$$

where the sums of the terms $O(\cdot)$ are denoted by dots.

For real ρ all the functions J are real-valued; therefore, for real $\rho > 0$, formulas (7.5.33) hold, where all the quantities are real.

If ρ is purely imaginary, $\arg\rho = \pi/2$, $\rho = i|\rho|$, we have

$$J_{\mu}(\xi) = i^{\mu}I_{\mu}(|\xi|), \qquad J_{-\mu}(\xi) = i^{-\mu}I_{-\mu}(|\xi|),$$
$$J_{1-\mu}(\xi) = i^{1-\mu}I_{1-\mu}(|\xi|), \qquad J_{-(1-\mu)}(\xi) = i^{-(1-\mu)}I_{-(1-\mu)}(|\xi|),$$

where the modified Bessel functions I have the real values.

This means, that for ρ purely imaginary, $\arg\rho = \pi/2$, $|\xi| \le N$, $|\rho| > \rho_N$ we have

$$\begin{aligned}
\varphi(x,\rho) &= \frac{\Gamma(1-\mu)}{2^{\mu}\Psi(0)}\,\Psi(x)|\xi|^{\mu} I_{-\mu}(|\xi|) + \dots,\\
\psi(x,\rho) &= \frac{\Gamma(\mu)\Psi(0)}{2^{1-\mu}}\,\Psi(x)|\rho|^{-2\mu}|\xi|^{\mu} I_{\mu}(|\xi|) + \dots,\\
\varphi'(x,\rho) &= \frac{\Gamma(1-\mu)}{2^{\mu}\Psi(0)}\,\frac{1}{\Psi(x)}\,|\rho|^{2\mu}|\xi|^{1-\mu} I_{(1-\mu)}(|\xi|) + \dots,\\
\psi'(x,\rho) &= \frac{\Gamma(\mu)\Psi(0)}{2^{1-\mu}}\,\frac{1}{\Psi(x)}\,|\xi|^{1-\mu} I_{-(1-\mu)}(|\xi|) + \dots,
\end{aligned}$$

where the sums of the terms $O(\cdot)$ are denoted by dots. However, in this section, we shall not need these formulas for ρ purely imaginary.

In order to obtain the desired estimates for $|k| \to \infty$ for the functions U_e, U_h we find at first the asymptotic estimates for each of the functions

$$A(k) = A_1(k) + ik\varkappa A_2(k), \tag{7.5.34}$$

where $A_1(k), A_2(k)$ are defined by formulas (7.5.5),

$$y_e(z_2,k) = -\{\varphi_e'(h_1,ik)\psi_e(z_2,ik) - \psi_e'(h_1,ik)\varphi_e(z_2,ik)\},\quad 0 \le z_2 \le h_1, \tag{7.5.35}$$

$$y_h(z,k) = A_1(k)\psi_h(z,k) - A_2(k)\varphi_h(z,k),\quad 0 \le z \le h, \tag{7.5.36}$$

by means of which the numerators and the denominators are expressed in formulas (7.5.4) for U_e, U_h.

We introduce the nonnegative functions $w_h(z)$, $w_e(z)$, $\tilde w_h(z_1)$, $\tilde w_e(z_2)$ by the formulas

$$\begin{aligned}
|K(z)| &= w_h^2(z), && 0 \le z \le h,\\
\tilde w_h(z_1) &= w_h(z), && z_1 = h - z,\quad 0 \le z_1 \le h,\\
K(z) &= w_e^2(z), && h \le z \le H,\\
\tilde w_e(z_2) &= w_e(z), && z_2 = z - h,\quad 0 \le z_2 \le h_1,
\end{aligned}$$

so that $\tilde K(z_2) = \tilde w_e^2(z_2)$ and set $\mu_i = 1/(\nu_i + 2)$, $i = 1, 2$.

7.5.2. Estimate for the function $A(k)$. Initial conditions (7.5.27) for the solutions $\varphi(x,\rho)$, $\psi(x,\rho)$ of equations (7.5.21) are set in the point $x = 0$, i. e., in the zero of the function $w^2(x)$. However, initial conditions

(7.5.10), (7.5.11) for the solutions $\varphi_h(z,k)$, $\psi_h(z,k)$ of equation (7.5.9) are set not in the zero of the coefficient $w_h^2(z)$. Therefore, in order to obtain the estimates for $\varphi_h(h,k)$, $\varphi_h'(h,k)$, $\psi_h(h,k)$, $\psi_h'(h,k)$ entering expression (7.5.5) for $A_1(k)$ and $A_2(k)$, we shall make some transformations. For this purpose, we express these four functions via solutions of type $\varphi(x,\rho)$, $\psi(x,\rho)$, to which formulas (7.5.30), (7.5.32), (7.5.33) can be applied.

Let $\tilde{\varphi}(x,k)$ and $\tilde{\psi}(x,k)$ be the solutions of the equation

$$y'' + k^2 w_h^2(z) y = 0, \qquad 0 \le z \le h, \tag{7.5.37}$$

with the initial conditions (in the point $z = h$)

$$\tilde{\varphi}(h,k) = 1, \quad \tilde{\varphi}'(h,k) = 0, \qquad \tilde{\psi}(h,k) = 0, \quad \tilde{\psi}'(h,k) = 1. \tag{7.5.38}$$

Express now the functions $\varphi_h(z,k)$, $\psi_h(z,k)$ via the functions $\tilde{\varphi}(z,k)$, $\tilde{\psi}(z,k)$. Since

$$\tilde{\psi}'(z,k)\tilde{\varphi}(z,k) - \tilde{\varphi}'(z,k)\tilde{\psi}(z,k) \equiv 1, \qquad 0 \le z \le h,$$

then, evidently, we obtain

$$\varphi_h(z,k) = \tilde{\psi}'(0,k)\tilde{\varphi}(z,k) - \tilde{\varphi}'(0,k)\tilde{\psi}(z,k),$$
$$\psi_h(z,k) = \tilde{\varphi}(0,k)\tilde{\psi}(z,k) - \tilde{\psi}(0,k)\tilde{\varphi}(z,k).$$

Therefore, we have

$$\varphi_h'(z,k) = \tilde{\psi}'(0,k)\tilde{\varphi}'(z,k) - \tilde{\varphi}'(0,k)\tilde{\psi}'(z,k),$$
$$\psi_h'(z,k) = \tilde{\varphi}(0,k)\tilde{\psi}'(z,k) - \tilde{\psi}(0,k)\tilde{\varphi}'(z,k).$$

Hence it follows that

$$\varphi_h(h,k) = \tilde{\psi}'(0,k), \qquad \varphi_h'(h,k) = -\tilde{\varphi}'(0,k),$$
$$\psi_h(h,k) = -\tilde{\psi}(0,k), \qquad \psi_h'(h,k) = \tilde{\varphi}(0,k). \tag{7.5.39}$$

We set now for $0 \le z \le h$ that

$$\tilde{w}_h(z_1) = w_h(z), \qquad z_1 = h - z, \quad 0 \le z_1 \le h_1,$$
$$\tilde{\varphi}(z,k) = \tilde{\varphi}_1(z_1,k), \qquad \tilde{\psi}(z,k) = -\tilde{\psi}_1(z_1,k). \tag{7.5.40}$$

The functions $\tilde{\varphi}_1(z_1,k)$ and $\tilde{\psi}_1(z_1,k)$, evidently satisfy the equation

$$y'' + k^2 \tilde{w}_h^2(z_1) y = 0, \qquad 0 \le z_1 \le h \tag{7.5.41}$$

and the initial conditions

$$\tilde{\varphi}_1(0,k) = 1, \quad \tilde{\varphi}_1'(0,k) = 0, \qquad \tilde{\psi}_1(0,k) = 0, \quad \tilde{\psi}_1'(0,k) = 1, \tag{7.5.42}$$

where the prime denotes the differentiation with respect to z_1.

Therefore, we can apply the formulas (7.5.30) to the functions $\tilde{\varphi}_1(z_1,k)$ and $\tilde{\psi}_1(z_1,k)$. In this case we should change

$$\begin{gathered}\rho \to k, \quad x \to z_1, \quad \mu \to \mu_1, \quad \xi \to \tilde{\xi}_1(z_1,k) = k\tilde{\Phi}(z_1),\\ \Phi(x) \to \tilde{\Phi}_1(z_1) = \int_0^{z_1} \tilde{w}_h(z_1)\,\mathrm{d}z_1,\\ \Psi(x) \to \tilde{\Psi}_1(z_1) = \{\tilde{\Phi}_1(z_1)\}^{1/2-\mu_1}/\{\tilde{w}_h(z_1)\}^{1/2}.\end{gathered} \tag{7.5.43}$$

Introduce the following functions (for $0 \le z \le h$)

$$\begin{gathered}\xi_1(z,k) = k\int_z^h w_h(z)\,\mathrm{d}z, \qquad \Phi_1(z) = \int_z^h w_h(z)\,\mathrm{d}z,\\ \Psi_1(z) = \{\Phi_1(z)\}^{1/2-\mu_1}/\{w_h(z)\}^{1/2}.\end{gathered} \tag{7.5.44}$$

Evidently, for $0 \le z \le h$, $z_1 = h - z$, we have

$$\tilde{\xi}_1(z_1,k) = \xi_1(z,k), \quad \tilde{\Phi}_1(z_1) = \Phi_1(z), \quad \tilde{\Psi}_1(z_1) = \Psi_1(z) \tag{7.5.45}$$

and hence

$$\tilde{\xi}_1(h,k) = \xi_1(0,k), \quad \tilde{\Psi}_1(0) = \Psi_1(h), \quad \tilde{\Psi}_1(h) = \Psi_1(0).$$

Taking into account equalities (7.5.39), (7.5.40), we obtain

$$\begin{aligned}\varphi_h(h,k) &= \tilde{\psi}'(0,k) = \tilde{\psi}_1'(h,k), \quad \psi_h(h,k) = -\tilde{\psi}(0,k) = \tilde{\psi}_1(h,k),\\ \varphi_h'(h,k) &= -\tilde{\varphi}'(0,k) = \tilde{\varphi}_1'(h,k), \quad \psi_h'(h,k) = \tilde{\varphi}(0,k) = \tilde{\varphi}_1(h,k).\end{aligned}$$

Introduce the constants

$$C_\varphi(\mu) = \Gamma(1-\mu)/2^\mu, \qquad C_\psi(\mu) = \Gamma(\mu)/2^{1-\mu}. \tag{7.5.46}$$

Formulas (7.5.30) yield the following statement.

There exist such numbers $N > 0$ and $k_N > 0$, that for real k $(k > 0)$, $\tilde{\xi}_1(h,k) = \xi_1(0,k) > N$, $k > k_N$ the following formulas hold

$$\varphi_h(h,k) = -C_\psi(\mu_1)\,\frac{\Psi_1(h)}{\Psi_1(0)}\,\xi_1(0,k)^{1/2-\mu_1}\sin\Big\{\xi_1(0,k) - \frac{\pi}{2}\Big(\frac{1}{2}+\mu_1\Big)\Big\} + \dots,$$

$$\psi_h(h,k) = C_\psi(\mu_1)\Psi_1(0)\Psi_1(h)k^{-2\mu_1}\xi_1(0,k)^{\mu_1-1/2} \times \cos\Big\{\xi_1(0,k) - \frac{\pi}{2}\Big(\frac{1}{2}+\mu_1\Big)\Big\} + \dots,$$

$$\varphi_h'(h,k) = -C_\varphi(\mu_1)\,\frac{1}{\Psi_1(0)\Psi_1(h)}\,k^{2\mu_1}\xi_1(0,k)^{1/2-\mu_1} \times \sin\Big\{\xi_1(0,k) - \frac{\pi}{2}\Big(\frac{1}{2}-\mu_1\Big)\Big\} + \dots,$$

$$\psi_h'(h,k) = C_\varphi(\mu_1)\,\frac{\Psi_1(0)}{\Psi_1(h)}\,\xi_1(0,k)^{\mu_1-1/2}\cos\Big\{\xi_1(0,k) - \frac{\pi}{2}\Big(\frac{1}{2}-\mu_1\Big)\Big\} + \dots. \tag{7.5.47}$$

Here and below we denote the sums of terms $O(\cdot)$, which have greater order of proximity than written terms, as dots.

Initial conditions (7.5.7), (7.5.8) for the solutions $\varphi_e(z_2,s)$, $\psi_e(z_2,s)$ of equation (7.5.6) are set in the zero $z_2 = 0$ of the coefficient $\tilde{w}_e^2(z_2) = \tilde{K}(z_2)$; therefore, we can apply directly formulas (7.5.32) to the functions $\varphi_e(z_2,s)$, $\psi_e(z_2,s)$. In this case we should replace

$$\rho \to s = ik, \quad x \to z_2, \quad \mu \to \mu_2, \quad \xi \to \tilde{\xi}_2(z_2,s) = s\tilde{\Phi}_2(z_2),$$
$$\Phi(x) \to \tilde{\Phi}_2(z_2) = \int_0^{z_2} \tilde{w}_e(z_2)\,\mathrm{d}z_2,$$
$$\Psi(x) \to \tilde{\Psi}_2(z_2) = \{\tilde{\Phi}_2(z_2)\}^{1/2-\mu_2}/\{\tilde{w}_e(z_2)\}^{1/2}. \tag{7.5.48}$$

Introduce the functions (for $h \le z \le H$)

$$\xi_2(z,s) = s\int_h^z w_e(z)\,\mathrm{d}z, \qquad \Phi_2(z) = \int_h^z w_e(z)\,\mathrm{d}z,$$
$$\Psi_2(z) = \{\Phi_2(z)\}^{1/2-\mu_2}/\{w_e(z)\}^{1/2}. \tag{7.5.49}$$

Evidently, for $h \le z \le H$, $z_2 = z - h$, we have

$$\tilde{\xi}_2(z_2,s) = \xi_2(z,s), \quad \tilde{\Phi}_2(z_2) = \Phi_2(z), \quad \tilde{\Psi}_2(z_2) = \Psi_2(z)$$

and

$$\tilde{\xi}_2(h_1,s) = \xi_2(H,s), \quad \tilde{\Psi}_2(0) = \Psi_2(h), \quad \tilde{\Psi}_2(h_1) = \Psi_2(H).$$

Therefore, formulas (7.5.32) yield the following statement.

There exist such numbers $N > 0$ and $k_N > 0$, that for real k ($k > 0$) (for purely imaginary $s = ik$, $\operatorname{Im} s > 0$, $s = i|s|$) and for $|\xi_2(H,s)| > N$, $k > k_N$, the following formulas hold:

$$\varphi_e'(h_1, ik) = C_\varphi(\mu_2)\,\frac{1}{\Psi_2(h)\Psi_2(H)}\,|k|^{2\mu_2}|\xi_2(H,s)|^{1/2-\mu_2}e^{|\xi_2(H,s)|} + \dots,$$
$$\psi_e'(h_1, ik) = C_\psi(\mu_2)\,\frac{\Psi_2(h)}{\Psi_2(H)}\,|\xi_2(H,s)|^{1/2-\mu_2}e^{|\xi_2(H,s)|} + \dots. \qquad (7.5.50)$$

Formulas (7.5.5), (7.5.47), (7.5.50) yield the following asymptotic estimates for $A_1(k)$, $A_2(k)$.

There exist such numbers $N > 0$ and $k_N > 0$, that for real values k ($k > 0$), $\xi_1(0,k) > N$, $|\xi_2(H,s)| > N$ ($s = ik$), $k > k_N$ the functions $A_1(k)$, $A_2(k)$ can be written in the form

$$A_1(k) = -\delta_1\xi_1(0,k)^{1/2-\mu_1}|\xi_2(H,s)|^{1/2-\mu_2}e^{|\xi_2(H,s)|}$$
$$\times\{\alpha_1\sin(\gamma_1-\gamma_2)+\alpha_2\sin(\gamma_1+\gamma_2)\} + \dots,$$
$$kA_2(k) = \delta_2\xi_1(0,k)^{1/2-\mu_1}|\xi_2(H,s)|^{1/2-\mu_2}e^{|\xi_2(H,s)|}$$
$$\times\{\alpha_1\cos(\gamma_1-\gamma_2)+\alpha_2\cos(\gamma_1+\gamma_2)\} + \dots,$$

where

$$\delta_1 = \{\Psi_1(0)\Psi_2(H)\}^{-1}, \quad \delta_2 = \Psi_1(0)\{\Psi_2(H)\Phi_1(0)^{1-2\mu_1}\}^{-1},$$
$$\alpha_1 = C_1|k|^{2\mu_2}/\pi, \qquad \alpha_2 = C_2|k|^{2\mu_1}/\pi,$$
$$C_1 = C_\psi(\mu_1)C_\varphi(\mu_2)\Psi_1(h)/\Psi_2(h), \quad C_2 = C_\varphi(\mu_1)C_\psi(\mu_2)\Psi_2(h)/\Psi_1(h),$$
$$\gamma_1 = \xi_1(0,k) - \pi/4, \qquad \gamma_2 = \mu_1\pi/2, \qquad (7.5.51)$$

and the values $C_\varphi(\mu)$, $C_\psi(\mu)$ are defined by formulas (7.5.46).

Recall that ($s = ik$)

$$\xi_1(0,k) = k\int_0^h w_h(z)\,dz, \qquad \xi_2(H,s) = s\int_h^H w_e(z)\,dz,$$
$$\Phi_1(z) = \int_z^h w_h(z)\,dz, \qquad \Phi_2(z) = \int_h^z w_e(z)\,dz,$$
$$\Psi_1(z) = \{\Phi_1(z)\}^{1/2-\mu_1}/\{w_h(z)\}^{1/2}, \qquad \Psi_2(z) = \{\Phi_2(z)\}^{1/2-\mu_2}/\{w_e(z)\}^{1/2},$$
$$|K(z)| = w_h^2(z), \qquad 0 \le z \le h; \qquad K(z) = w_e^2(z), \quad h \le z \le H. \quad (7.5.52)$$

Hence, for the function $A(k)$ of the form (7.5.34) we obtain the expression

$$A(k) = P(k)\{-\delta_1\{\alpha_1\sin(\gamma_1-\gamma_2)+\alpha_2\sin(\gamma_1+\gamma_2)\}$$
$$+ i\varkappa\delta_2\{\alpha_1\cos(\gamma_1-\gamma_2)+\alpha_2\cos(\gamma_1+\gamma_2)\}\} + \dots$$

where

$$P(k) = \xi_1(0,k)^{1/2-\mu_1} |\xi_2(H,s)|^{1/2-\mu_2} e^{|\xi_2(H,s)|}. \qquad (7.5.53)$$

Since the functions $\Psi_j(z)$ and $\{\Psi_j(z)\}^{-1}$ are bounded (taking into account (7.5.23)), $\mu_1 \le 1/2$ then the numbers δ_j, α_j ($j = 1, 2$) are finite.

Therefore,

$$|A(k)| \ge Q \min(\delta_1, \varkappa\delta_2) |P(k)| R(k),$$

where Q is a certain positive constant independent of k; $\min(\delta_1, \varkappa\delta_2)$ is the minimal number from δ_1 and $\varkappa\delta_2$;

$$R(k) = \{\alpha_1^2 + \alpha_2^2 + 2\alpha_1\alpha_2 \cos\{(\gamma_1+\gamma_2) - (\gamma_1-\gamma_2)\}\}^{1/2}$$
$$= \{\alpha_1^2 + \alpha_2^2 + 2\alpha_1\alpha_2 \cos(2\gamma_2)\}^{1/2}.$$

Since $0 < 2\gamma_2 = \mu_1\pi \le \pi/2$, then $0 \le \cos(2\gamma_2) < 1$; therefore,

$$R(k) \ge \{\alpha_1^2 + \alpha_2^2\}^{1/2} > 0.$$

As a result we obtain the following estimate for the function $A(k)$.

There exist such numbers $N > 0$ and $k_0 > 0$, that for real values k, $-\infty < k < \infty$ and for $\xi_1(0,k) > N$, $|\xi_2(H,s)| > N$, $|k| > k_0$ the following inequality holds:

$$|A(k)| > Q|P(k)| \{[C_1|k|^{2\mu_2}]^2 + [C_2|k|^{2\mu_1}]^2\}^{1/2}. \qquad (7.5.54)$$

Here Q is a certain positive constant independent of k, and the numbers C_1, C_2 and the function $P(k)$ is defined as in (7.5.51), (7.5.52), (7.5.53).

7.5.3. Estimates for the functions $y_e(z_2,k)$, $U_e(z_2,k)$. Note that the function $y_e(z_2,k)$, defined by formula (7.5.35) satisfies the equation

$$y'' - k^2\tilde{K}(z_2)y = 0, \qquad 0 \le z_2 \le h_1 \qquad (7.5.55)$$

and the initial conditions

$$y_e(h_1,k) = 1, \qquad y_e'(h_1,k) = 0$$

(the prime denotes the differentiation with respect to z_2).

We have $\tilde{K}(z_2) \ge 0$. Therefore, taking into account the properties of solutions of the Sturm—Liouville equation (7.5.55), we can say that the functions $y_e(z_2,k)$, $y_e'(z_2,k)$ do not increase in the segment $0 \le z_2 \le h_1$.

So, we can write

$$y_e(z_2,k) \le y_e(0,k) = \psi_e'(h_1,ik), \qquad y_e'(z_2,k) \le y_e'(0,k) = -\varphi_e'(h_1,ik).$$

For the functions $\psi_e'(h_1,ik)$, $\varphi_e'(h_1,ik)$, we have formulas (7.5.50). Applying these formulas, we obtain the following estimates.

There exist the numbers $N>0$ and $k_N>0$, such that for real k ($k>0$) and for $|\xi_2(H,s)|>N$, $s=ik$, $k>k_N$, $0\le z_2\le h_1$, the following inequalities hold

$$\begin{aligned} y_e(z_2,k) &\le C_\psi(\mu_2)\,\frac{\Psi_2(h)}{\Psi_2(H)}\,|\xi_2(H,s)|^{1/2-\mu_2}e^{|\xi_2(H,s)|}+\ldots,\\ |y_e'(z_2,k)| &\le C_\varphi(\mu_2)\,\frac{1}{\Psi_2(h)\Psi_2(H)}\,|k|^{2\mu_2}|\xi_2(H,s)|^{1/2-\mu_2}e^{|\xi_2(H,s)|}+\ldots. \end{aligned} \tag{7.5.56}$$

The values C_φ, C_ψ, $\Psi_2(z)$, $\xi_2(H,s)$ entering these inequalities are expressed by the formulas (7.5.46), (7.5.52). The formula

$$U_e(z_2,k) = -ik\Phi(k)y_e(z_2,k)/A(k),$$

equalities (7.5.20), and estimates (7.5.54), (7.5.56) yield the following statement.

There exist such $k_0>0$, that for $|k|>k_0$, $-\infty<k<\infty$, $0\le z_2\le h$, we have

$$\begin{aligned} |U_e(z_2,k)| &< |\Phi(k)|\,|k|\,\frac{Q_0|k|^{\mu_1-1/2}}{\{[C_1|k|^{2\mu_2}]^2+[C_2|k|^{2\mu_1}]^2\}^{1/2}},\\ \left|\frac{\mathrm{d}U_e(z_2,k)}{\mathrm{d}z_2}\right| &< |\Phi(k)|\,|k|\,\frac{Q_1|k|^{2\mu_2}|k|^{\mu_1-1/2}}{\{[C_1|k|^{2\mu_2}]^2+[C_2|k|^{2\mu_1}]^2\}^{1/2}}, \end{aligned}$$

where the constants Q_0, Q_1, C_1, C_2 are independent of $z_2\in[0,h_1]$. The constants C_1, C_2 are defined by formulas (7.5.51), and $\mu_i = 1/(\nu_i+2)$, $i=1,2$.

Hence it follows that for $|k|>k_0$, $-\infty<k<\infty$, $0\le z_2\le h_1$ the inequalities ($n=0,1,2$)

$$\left|\frac{\mathrm{d}^nU_e(z_2,k)}{\mathrm{d}z_2^n}\right| < Q_n|\Phi(k)||k|^{\delta_e(n)}|k|^{1/2+\mu_1-2\max(\mu_1,\mu_2)}, \tag{7.5.57}$$

hold, where

$$\delta_e(n) = \begin{cases} 2\mu_2 n & \text{for} \quad n=0,1,\\ 2 & \text{for} \quad n=2. \end{cases}$$

The constants Q_n here are independent of $z_2 \in [0, h_1]$. Take into account the condition (B) of Theorem 7.5.1 and the connection between the smoothness of the function $\varphi(\xi)$ and the speed of decreasing for $|k| \to \infty$ of its Fourier transform $\Phi(k)$. As a result we have that there exists such $k_1 > 0$ that for $|k| > k_1$, $-\infty < k < \infty$, $0 \le z_2 \le h_1$, $n = 0, 1, 2$ the following estimates hold

$$\left| \frac{\mathrm{d}^n U_e(z_2, k)}{\mathrm{d}z_2^n} \right| < Q_e |k|^{\delta_e(n)} |k|^{-7/2+\mu_1-2\max(\mu_1,\mu_2)}. \tag{7.5.58}$$

7.5.4. Estimates for the functions $y_h(z, k)$, $U_h(z, k)$. First we note that the function $y_h(z, k)$ defined by formula (7.5.36) satisfies equation (7.5.37) and the initial conditions

$$y_h(h, k) = -\psi_e'(h_1, ik), \qquad y_h'(h, k) = \varphi_e'(h_1, ik).$$

Therefore, the function $y_h(z, k)$ can be represented as follows:

$$y_h(z, k) = -\psi_e'(h_1, ik)\tilde{\varphi}(z, k) + \varphi_e'(h_1, ik)\tilde{\psi}(z, k), \tag{7.5.59}$$

where $\tilde{\varphi}(z, k)$, $\tilde{\psi}(z, k)$ solve equation (7.5.37) for the initial conditions (7.5.38). As was above noted, if, for $0 \le z \le h$, we set $z_1 = h - z$, $0 \le z_1 \le h_1$, then

$$\tilde{\varphi}(z, k) = \tilde{\varphi}_1(z_1, k), \quad \tilde{\psi}(z, k) = -\tilde{\psi}_1(z_1, k), \tag{7.5.60}$$

where the functions $\tilde{\varphi}_1(z_1, k)$ and $\tilde{\psi}_1(z_1, k)$ satisfies equation (7.5.41) and initial conditions (7.5.42). Taking into account formulas (7.5.43)–(7.5.45), we apply formulas (7.5.30) to the functions $\tilde{\varphi}_1(z_1, k)$ and $\tilde{\psi}_1(z_1, k)$. As a result we obtain, that for $|\xi_1(z, k)| > N$, $k > k_N$,

$$\begin{aligned}
\tilde{\varphi}_1(z_1, k) &= C_\varphi(\mu_1) \frac{\Psi_1(z)}{\Psi_1(h)} \xi_1(z, k)^{\mu_1 - 1/2} \cos\left\{\xi_1(z, k) - \frac{\pi}{2}\Big(\frac{1}{2} - \mu_1\Big)\right\} + \ldots, \\
\tilde{\psi}_1(z_1, k) &= C_\psi(\mu_1)\Psi_1(h)\Psi_1(z)|k|^{-2\mu_1}\xi_1(z, k)^{\mu_1 - 1/2} \\
&\quad \times \cos\left\{\xi_1(z, k) - \frac{\pi}{2}\Big(\frac{1}{2} + \mu_1\Big)\right\} + \ldots, \\
\tilde{\varphi}_1'(z_1, k) &= -C_\varphi(\mu_1) \frac{1}{\Psi_1(h)\Psi_1(z)} \rho^{2\mu_1} \xi_1(z, k)^{1/2 - \mu_1} \\
&\quad \times \sin\left\{\xi_1(z, k) - \frac{\pi}{2}\Big(\frac{1}{2} - \mu_1\Big)\right\} + \ldots, \\
\tilde{\psi}_1'(z_1, k) &= -C_\psi(\mu_1) \frac{\Psi_1(h)}{\Psi_1(z)} \xi_1(z, k)^{1/2 - \mu_1} \sin\left\{\xi_1(z, k) - \frac{\pi}{2}\Big(\frac{1}{2} + \mu_1\Big)\right\} + \ldots.
\end{aligned} \tag{7.5.61}$$

Since the functions $\Psi_1(z)$ and $\{\Psi_1(z)\}^{-1}$ are bounded from below and from above (this follows from (7.5.23)) in the segment $0 \le z \le h$ and $|\sin(\cdot)| \le 1$, $|\cos(\cdot)| \le 1$, $\xi_1(z,k) \le \xi_1(0,k)$, then, taking into account (7.5.61), we obtain the estimates

$$|\tilde{\varphi}(z,k)| \le \tilde{C}_1 \xi_1(z,k)^{\mu_1 - 1/2}, \quad |\tilde{\psi}(z,k)| \le \tilde{C}_2 |k|^{-2\mu_1} \xi_1(z,k)^{\mu_1 - 1/2},$$
$$|\tilde{\varphi}'(z,k)| \le \tilde{C}_3 |k|^{2\mu_1} \xi_1(0,k)^{1/2-\mu_1}, \quad |\tilde{\psi}'(z,k)| \le \tilde{C}_4 \xi_1(0,k)^{1/2-\mu_1} \qquad (7.5.62)$$

for $k > k_N$, $\xi_1(z,k) > N$, $0 \le z \le h$. Here $\tilde{C}_i$ $(i = 1,2,3,4)$ are certain positive constants.

The values $\varphi_e'(h_1, ik)$, $\psi_e'(h_1, ik)$ entering expression (7.5.36) satisfy the estimates (7.5.50).

Estimates (7.5.62), (7.5.50) yield the following statement.

There exist such numbers $N > 0$ and $k_N > 0$, that for real k $(k > 0)$ and for $\xi_1(z,k) > N$, $|\xi_2(H,s)| > N$, $k > k_N$, the following inequalities hold

$$|y_h(z,k)| < C|\xi_2(H,s)|^{1/2-\mu_2} \xi_1(z,k)^{\mu_1 - 1/2} e^{|\xi_2(H,s)|} \{D_1 + D_2 |k|^{2(\mu_2 - \mu_1)}\},$$
$$|y_h'(z,k)| < C|\xi_1(0,k)|^{1/2-\mu_1} |\xi_2(H,s)|^{1/2-\mu_2} e^{|\xi_2(H,s)|} \{\tilde{D}_1 |k|^{2\mu_2} + \tilde{D}_2 |k|^{2\mu_1}\}, \qquad (7.5.63)$$

where the constants C, D_n, $\tilde{D}_n$ $(n = 1,2)$ are independent of k, z.

Consider now the case when $|\xi_1(z,k)| \le N$.

In this case we use the formulas (7.5.33) for the functions $\tilde{\varphi}_1(z_1,k)$, $\tilde{\psi}_1(z_1,k)$ and the change (7.5.43)–(7.5.45). As a result we obtain for $|\xi_1(z,k)| \le N$, $|k| > k_N$, $0 \le z \le h$ the following formulas

$$\tilde{\varphi}_1(z_1,k) = C_\varphi(\mu_1) \frac{\Psi_1(z)}{\Psi_1(h)} \xi_1(z,k)^{\mu_1} J_{-\mu_1}(\xi_1(z,k)) + \dots,$$
$$\tilde{\psi}_1(z_1,k) = C_\psi(\mu_1) \Psi_1(h) \Psi_1(z) |k|^{-2\mu_1} \xi_1(z,k)^{\mu_1} J_{\mu_1}(\xi_1(z,k)) + \dots,$$
$$\tilde{\varphi}_1'(z_1,k) = -C_\varphi(\mu_1) \frac{1}{\Psi_1(h)\Psi_1(z)} |k|^{2\mu_1} \xi_1(z,k)^{1-\mu_1} J_{1-\mu_1}(\xi_1(z,k)) + \dots,$$
$$\tilde{\psi}_1'(z_1,k) = C_\psi(\mu_1) \frac{\Psi_1(h)}{\Psi_1(z)} \xi_1(z,k)^{1-\mu_1} J_{-(1-\mu_1)}(\xi_1(z,k)) + \dots. \qquad (7.5.64)$$

Since $\xi_1(z,k) \le \xi_1(0,k)$, $|J_\nu(x)| <$ const and $|x^\nu J_{-\nu}(x)| <$ const for each $\nu > 0$, and taking into account equalities (7.5.60), (7.5.64), we have

$$|\tilde{\varphi}(z,k)| < \tilde{C}_1, \qquad |\tilde{\psi}(z,k)| < \tilde{C}_2 |k|^{-2\mu_1} \xi_1(z,k)^{\mu_1},$$
$$|\tilde{\varphi}'(z,k)| < \tilde{C}_3 |k|^{2\mu_1} \xi_1(z,k)^{1-\mu_1}, \qquad |\tilde{\psi}'(z,k)| < \tilde{C}_4, \qquad (7.5.65)$$

where $\tilde{C}_n$, $n = 1, 2, 3, 4$, are certain positive constants. Estimates (7.5.54), (7.5.65) yield the following statement.

There exist the numbers $N > 0$ and $k_N > 0$, such that for real k ($k > 0$) and for $|\xi_2(H, s)| > N$ ($s = ik$), $\xi_1(z, k) \leq N$, $k > k_N$, $0 \leq z \leq h$ the following inequalities hold

$$|y_h(z,k)| < C|\xi_2(H,s)|^{1/2-\mu_2} e^{|\xi_2(H,s)|}\{D_1 + D_2|k|^{2(\mu_2-\mu_1)}\xi_1(z,k)^{\mu_1}\},$$
$$|y_h'(z,k)| < C|\xi_2(H,s)|^{1/2-\mu_2} e^{|\xi_2(H,s)|}\{\tilde{D}_1|k|^{2\mu_2} + \tilde{D}_2|k|^{2\mu_1}\xi_1(z,k)^{1-\mu_1}\}, \tag{7.5.66}$$

where the constants C, D_n, $\tilde{D}_n$ ($n = 1, 2$) are independent of k, z.

Since $U_h(z, k) = ik\Phi(k)y_h(z, k)/A(k)$, then, inequalities (7.5.54), (7.5.63) yield the following estimates for the function $U_h(z, k)$.

There exist the number $N > 0$, $k_0 > 0$, such that for

$$\xi_1(z,k) = k\int_z^h w_h(z)\,\mathrm{d}z > N, \qquad |k| > k_0, \quad -\infty < k < \infty,$$

we have

$$|U_h(z,k)| < Q_0|\Phi(k)|\,|k|\,\frac{|k|^{-(\mu_1+1/2)}\{D_1|k|^{2\mu_1} + D_2|k|^{2\mu_2}\}}{\{[C_1|k|^{2\mu_2}]^2 + [C_2|k|^{2\mu_1}]^2\}^{1/2}},$$
$$\left|\frac{\mathrm{d}U_h(z,k)}{\mathrm{d}z}\right| < Q_1|\Phi(k)|\,|k|\,\frac{\{\tilde{D}_1|k|^{2\mu_1} + \tilde{D}_2|k|^{2\mu_2}\}}{\{[C_1|k|^{2\mu_2}]^2 + [C_2|k|^{2\mu_1}]^2\}^{1/2}},$$

where the constants Q_0, Q_1, C_1, C_2, D_1, D_2, $\tilde{D}_1$, $\tilde{D}_2$ are independent of k, z.

Therefore, for such values $\xi_1(z, k)$, z, k we have

$$\left|\frac{\mathrm{d}^n U_h(z,k)}{\mathrm{d}z^n}\right| < Q_n|\Phi(k)|\,|k|^{\delta_h(n)}|k|^{1/2-\mu_1}, \tag{7.5.67}$$

where

$$\delta_h(n) = \begin{cases} (1/2+\mu_1)n & \text{for} \quad n = 0, 1, \\ 2 & \text{for} \quad n = 2. \end{cases}$$

Taking into account inequalities (7.5.54), (7.5.66), we obtain that for the same values $N > 0$ and $k_0 > 0$ and for $\xi_1(z, k) \leq N$, $|k| > k_0$, $-\infty < k < \infty$, the following inequalities hold

$$|U_h(z,k)| < Q_0|\Phi(k)|\,|k|\,\frac{|k|^{-(\mu_1+1/2)}\{D_1|k|^{2\mu_1} + D_2|k|^{2\mu_2}\}}{\{[C_1|k|^{2\mu_2}]^2 + [C_2|k|^{2\mu_1}]^2\}^{1/2}},$$

(this is the estimate similar to those for $\xi_1(z,k) > N$)

$$\left|\frac{\mathrm{d}U_h(z,k)}{\mathrm{d}z}\right| < Q_1|\Phi(k)|\,|k|\,\frac{|k|^{\mu_1-1/2}\{\tilde{D}_1|k|^{2\mu_1}+\tilde{D}_2|k|^{2\mu_2}\}}{\{[C_1|k|^{2\mu_2}]^2+[C_2|k|^{2\mu_1}]^2\}^{1/2}},$$

where the constants Q_0, Q_1, C_1, C_2, D_1, D_2, $\tilde{D}_1$, $\tilde{D}_2$ are independent of k, z.

Since $|k|^{\mu_1-1/2} < 1$ for $|k| > k_0 > 1$, then the following final estimates hold.

There exist such numbers $k_0 > 0$ and $Q_n > 0$, independent of z, so that for $|k| \geq k_0$, $-\infty < k < \infty$, $0 \leq z \leq h$ the inequalities of form (7.5.67) hold ($n = 0, 1, 2$). Or, under the condition (B) of Theorem 7.5.1, the following inequalities hold:

$$\left|\frac{\mathrm{d}^n U_h(z,k)}{\mathrm{d}z^n}\right| < Q_h|k|^{\delta_h(n)}|k|^{-7/2-\mu_1}, \qquad n = 0, 1, 2. \tag{7.5.68}$$

Combining the estimates obtained for the functions U_e, U_h we can formulate the following lemma.

Lemma 7.5.2. *Under the conditions of Theorem 7.5.1, the functions $U_e(z_2,k)$, $U_h(z,k)$ satisfy the estimates* (7.5.57), (7.5.58), (7.5.67), (7.5.68) *for $|k| \to \infty$, $-\infty < k < \infty$, $0 \leq z_2 \leq h_1$, $0 \leq z \leq h$.*

From estimate (7.5.58), (7.5.68) it follows that each of the functions $k^n U_e(z_2,k)$, $U_e^{(n)}(z_2,k)$, $k^n U_h(z,k)$, $U_h^{(n)}(z,k)$ ($n = 0,1,2$) is absolutely integrable in the line $-\infty < k < \infty$. Therefore, it can be represented as the Fourier integral (as a complex-valued function of real variable k) for each $z_2 \in [0, h_1]$, $z \in [0, h]$, and the integral in formula (7.5.2) admits twice the differentiation under the integral sign with respect to z and ξ. Since $u(z,\xi)$ is the Fourier transform of the function $U(z,k)$ which is represented by the Fourier integral, then the integral in equality (7.5.2) is the Fourier integral (just as the integrals for the derivatives u_z, u_ξ, u_{zz}, $u_{\xi\xi}$, $u_{z\xi}$).

From equalities (7.5.20) it follows that all these integrals are real.

The properties of the functions $U_e(z_2,k)$ and $U_h(z,k)$ show that the function $u(z,\xi)$ defined by formulas (7.5.2)–(7.5.11) satisfies all the conditions of the direct problem 7.1 (condition (7.1.7) holds because $u(z,\xi) \to 0$, for $|\xi| \to \infty$, uniformly with respect to z in the segment $[0, H]$. This follows from estimates (7.5.58), (7.5.68) and the Riemann—Lebesgue theorem). Theorem 7.5.1 is proved. □

We have proved that the solution of the direct problem exists if the conditions of Theorem 7.5.1 hold. Hence, by Theorem 7.1.1, we obtain that the solution of the direct problem is unique and coincides with the constructed solution if the conditions of Theorem 7.5.1 hold. [v)]

7.5.5. Inverse problems in the general case. Formulations of inverse problems 7.1–7.5 for the general case are just the same that those in § 4 formulated for the case $K(h+0) \neq 0$, $K(h-0) \neq 0$. The method of solution of the unique theorem for all inverse problems is common. All these problems are reduced to the well-known inverse problem for the one-dimensional equation for the regular string

$$y'' + \lambda \rho(z) y = 0$$

in the segment $0 \leq z \leq L < \infty$, $\rho(z) \geq 0$; λ the spectral parameter. In this problem we need to find the nonnegative density $\rho(z) \geq 0$ and the length L of the string given its spectral function or the main transition function. In this case we use essentially the representation of the solution of the direct problem given by Theorem 7.5.1. This method uses the definition of the function

$$\Omega(s) = \psi_e'(h_1, s)/\varphi_e'(h_1, s).$$

We shall demonstrate now this method taking the inverse problem 7.1. Taking into account Theorem 7.5.1, we have

$$U(H,k) = -\frac{ik\Phi(k)}{A_1(k) + ik\varkappa A_2(k)}.$$

Inverting integral (7.5.2) for $z = H$ and using the function $\varphi(\xi)$, define the function

$$A(k) = A_1(k) + ik\varkappa A_2(k)$$

in a certain segment $[a, b]$ of the real axis. The number $\varkappa$ and the functions $A_1(k)$, $A_2(k)$ are real for $-\infty < k < \infty$ and $\varkappa \neq 0$. Therefore, separating real and imaginary parts of the function $A(k)$, we find the functions $A_1(k)$, $\varkappa A_2(k)$ in the segment $[a, b]$. Using the property $A_2(0) = 1$, find the number $\varkappa$ and the function $A_2(k)$ in the segment $[a, b]$.

Since the number h and the function $K(z)$ for $z \in [0, h]$ are known, we can find the functions $\varphi_h(z,k)$, $\psi_h(z,k)$ for $k \in [a, b]$. After, we take into account the identities

---

[v)] Also, we take into account that the functions $U_e^{(n)}$, $U_h^{(n)}$ $(n = 0, 1, 2)$ are differentiable with respect to k in the axis $-\infty < k < \infty$ for each fixed $z_2 \in [0, h_1]$, $z \in [0, h]$ (except, may be, the point $k = 0$) since the condition (C) holds and taking into account the analytical properties of the functions $\psi_e, \varphi_e, \varphi_h, \psi_h$.

$$A_1(k)\psi_h(h,k) - A_2(k)\varphi_h(h,k) = -\psi_e'(h_1, ik),$$
$$A_1(k)\psi_h'(h,k) - A_2(k)\varphi_h'(h,k) = \varphi_e'(h_1, ik),$$

and find the functions $\psi_e'(h_1, s)$ and $\varphi_e'(h_1, s)$ in the segment $[ia, ib]$ of the imaginary axis of the complex plane $s = ik$. Extend the entire functions $\psi_e'(h_1, h)$ and $\varphi_e'(h_1, s)$ (using, for example, the Taylor expansion) on the neighborhood of the real axis $-\infty < s < \infty$. As a result, we find the function $\Omega(s)$ in this neighborhood.

As is known (see Kats and Krein, 1964, § 3), if $\Omega(s)$ is known, then the spectral function $\tau(\lambda)$ of the regular string with equation

$$y'' + \lambda\tilde{K}(z_1)y = 0, \qquad \lambda = s^2, \quad 0 \le z_1 \le h_1$$

is constructed uniquely. Here $\rho(z_1) = \tilde{K}(z_1)$ is the density, equal to zero, may be, only in one point $z_1 = 0$; h_1 is the length of the string. The function $\Omega(s)$ is called the coefficient of dynamical pliability of the string (Kats and Krein, 1964, § 10). From the known unique theorems for the case of regular string (Krein, 1951a, 1951b, 1952, 1953; Kats and Krein, 1964, § 11) it follows that both $\Omega(s)$ and the spectral function $\tau(\lambda)$ (or the main transition function $\Phi(t)$ defined by $\tau(\lambda)$) define the density of the string $\rho(z_1)$ and its length h_1 uniquely. In order to find $\rho(z_1)$ and h_1 we apply the Krein method (see, Krein, 1951a,b, 1952, 1953). The number H we define from the formula $H = h_1 + h$.

The other inverse problems 7.2–7.4 are considered analogously; so, Theorem 7.4.1 holds in the general case.

7.6. THE OTHER PROBLEMS

7.6.1. The problems with the condition $u(H,\xi) = 0$. Change the condition (7.1.1) by the condition

$$u(H,\xi) = 0, \qquad -\infty < \xi < \infty \tag{7.1.1°}$$

in the direct problem 7.1. We shall call this problem the direct problem 7.2. Analogously Theorem 7.1.1 the following theorem holds.

Theorem 7.6.1. *Suppose that there exist two solutions of the direct problem 7.0.1,* (7.1.1°), (7.1.2)–(7.1.6). *Then their difference is a constant.*

All the results of § 7.2–7.5 hold and for condition (7.1.1°). For brevity we confine ourselves by the case $K(h+0) \neq 0$, $K(h-0) \neq 0$.

Using the method of § 7.2 we can establish the following theorem:

Theorem 7.6.2. *Let all the conditions of Theorem 7.2.1 hold relatively the direct problem 7.2. Then the solution $u_0(z,\xi)$ of the direct problem 7.2 exists, is unique, and has the representation which is obtained from formulas (7.2.1)–(7.2.15) by the change of the functions $\Phi_e(ik)$, $\Psi_e(ik)$ by the functions $\varphi_e(x_H, ik)$, $\psi_e(x_H, ik)$ respectively.*

Inverse problem 7.1°. Let all the conditions of Theorem 7.6.2 hold and the function $f(\xi)$ and the values $\partial u_0/\partial z|_{z=H}$ of the solution $u_0(z,\xi)$ of the direct problem 7.2 are set for $-\infty < \xi < \infty$. The function $K(z)$ is unknown in the segment $[h, H]$. One of the numbers $H, \varkappa$ is unknown; the other is set. Besides, the number h and the function $K(z)$ in the segment $[0, h]$ are set. We need to find those of the numbers $H, \varkappa$ that is unknown and the function $K(z)$ for $h \le z \le H$.

Inverse problem 7.2°, 7.3°, 7.4°, 7.5°. Replace the words "direct problem 7.1" by the words "direct problem 7.2" in the formulations of the inverse problems 7.2, 7.3, 7.4, 7.5. The formulations obtained as the result we shall call by the inverse problems 7.2°, 7.3°, 7.4° 7.5° respectively.

Theorem 7.6.3. *Each of the inverse problems 7.1°–7.5° has a unique solution.*

Solution of these problems and the proof of Theorem 7.6.3 follow from the methods of Subsections 7.4.1–7.4.3 and is based on the definition of the spectra of the problems obtained from the problems (I) and (II) § 7.4 by replacing the condition $\cos\theta y(x_H) + \sin\theta y'(x_H) = 0$, $\operatorname{ctg}\theta = -\overset{+}{\alpha}$ by the condition $y(x_H) = 0$.

7.6.2. Simultaneous set of the information on the solutions of the direct problem 7.1 and 7.2. Setting simultaneously the data on solution of the direct problems 7.1 and 7.2, we can formulate the inverse problems on determination of the coefficient $K(z)$ everywhere in the segment $[0, H]$. Consider, for example, the following setting.

Inverse problem 7.6. Let all the conditions of Theorem 7.2.1 which are formulated relatively the direct problems 7.1 and 7.2 hold. In these problems the numbers $H, h, \varkappa$ and the functions $K(z)$ coincide in each interval $[0, h]$, $[h, H]$. The functions $f(\xi)$ may not coincide. Let, besides, the values $u(h,\xi)$, $\partial u/\partial z|_{z=h}$ of the solution $u(z,\xi)$ of the direct problem 7.1 are set for $-\infty < \xi < \infty$ and the values $u_0(h,\xi)$ of the solution $u_0(z,\xi)$ of the direct problem 7.2 are set also. The numbers $H, h, \varkappa$ and the function $K(z)$ in each of the segments $[0, h]$ and $[h, H]$ are unknown. The function $f(\xi)$ is set for $-\infty < \xi < \infty$ for each of the direct problems 7.1 and 7.2.

We need to find the numbers H, h, $\varkappa$ and the function $K(z)$ in each of the segments $[0, h]$ and $[h, H]$.

First we define the spectra of the problems (I), (II) § 7.4. Then, we find the numbers x_H, $H - h$, $\varkappa$ and the function $\overset{+}{v}(x_1)$ for $0 \leq x_1 \leq x_H$. Finally we find the spectra of the problems

$$-y'' + \bar{q}(x)y = \lambda y, \qquad 0 \leq x \leq x_h,$$
$$\cos\varphi y(0) + \sin\varphi y'(0) = 0, \qquad \cos\theta y(x_h) + \sin\theta y'(x_h) = 0;$$

$$-y'' + \bar{q}(x)y = \lambda y, \qquad 0 \leq x \leq x_h,$$
$$y(0) = 0, \qquad \cos\theta y(x_h) + \sin\theta y'(x_h) = 0,$$
$$(\operatorname{ctg}\varphi = -\bar{\alpha}_0, \qquad \operatorname{ctg}\theta = -\bar{\alpha})$$

(the squares of zeros of the functions $\Phi_h(k)$ and $\Psi_h(k)$) and all the desired values and functions. As a result we establish the uniqueness of solution of the inverse problem 7.6.

7.6.3. Permutation of the domains of hyperbolicity and ellipticity. All the results stated in § 7.1–7.6 are transferred for the case when we change the domains of hyperbolicity and ellipticity of equation (7.0.1) by places. This means that in equation (7.0.1) we have

$$K(z) > 0 \quad \text{for} \quad z \in [0, h), \qquad K(z) < 0 \quad \text{for} \quad x \in (h, H]. \tag{7.6.1}$$

All other conditions of the direct problems 7.1 or 7.2 are concerved. We have taken about the uniqueness theorem for the direct problem 7.1 in the end of § 7.1. For example, in the case $K(h+0) \neq 0$, $K(h-0) \neq 0$, we can obtain the representation of solution of the direct problem with the coefficient of type (7.6.1) from Theorem 7.2.1. For this purpose we permutate everywhere in the formulas of this theorem e and h; $\bar{v}$ and $\overset{+}{v}$, $\bar{q}$ and $\overset{+}{g}$. The inverse problems in the case (7.6.1) are formulated and considered analogously to § 7.4–7.6.

7.7. THE PHYSICAL CONTENT

The direct and inverse problems for equation of the mixed type (7.0.1) considered above have the following physical sense and contents. As it follows

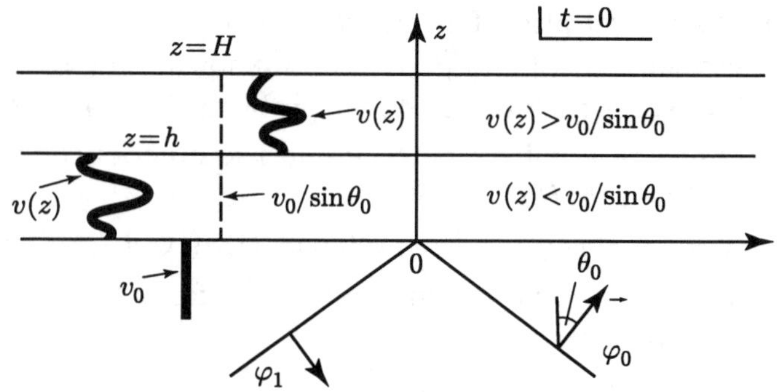

Figure 7.2.

from § 3.1, the direct problems 7.1 and 7.2 describe the vibration process in the inhomogeneous elastic layer of the depth H with a free of fixed boundary $z = H$, excited by the plane wave SH in the case $\mu(z) \equiv \mu_0 = \text{const}$. Here for $0 \leq z \leq h$ we have $v(z) < v_0/\sin\theta_0$ (this condition means that the apparent speed is greater than the speed of the wave propagation in the layer; this is the subcritical incidence of the plane wave) and for $h \leq z \leq H$ we have $v(z) > v_0/\sin\theta_0$ (here, the apparent speed is less than the speed in the layer; this is the supercritical incidence of the plane wave. It corresponds to the case of complete internal reflection). Following Subsection 7.6.3 these two conditions admit the permutation.

Here we have

$$K(z) = \sin^2\theta_0/v_0^2 - 1/v^2(z), \qquad 0 \leq z \leq H,$$
$$\varphi(\xi) = -2\varkappa\varphi_0(\xi), \qquad \varkappa = \cos\theta_0/v_0.$$

Here $v(z)$ and $\mu(z)$ are the speed of the wave propagation and the shift modulus respectively in the inhomogeneous layer $0 \leq z \leq H$; v_0 and μ_0 are the speed and the shift modulus respectively in the homogeneous half-space $z < 0$, whence the plane wave of the form $\varphi_0(\xi)$ is falling at the angle $\theta_0 \in (0, \pi/2)$.

The summary shift field in the layer is

$$u(z,\xi) = u(x,z,t) \quad (\xi = t - x\sin\theta_0/v_0),$$

where $u(0,\xi) = \varphi_0(\xi) + \varphi_1(\xi)$ and $\varphi_1(\xi)$ is the form of the plane wave reflected from the layer. The choice of the coordinate system x, z, t (t is the time) is shown in Figure 7.2.

Therefore, the inverse problems corresponds to the problems of determination of the characteristics of the inhomogeneous layer given the vibrations of a point from any boundary of the layer or given the forms of the incident and reflected plane waves. The function $v(z)$ may be not monotone.

Corresponding to Remark 3.5.2, the analogous physical sense of the direct and inverse problems of this chapter we obtain the acoustic case also.

All these problems have also the following physical interpretation. Equation of the mixed type (7.0.1) can be considered as the equation of the string vibration whose density can be both positive and negative in the various segments of the string. Respectively, the inverse problems for the mixed type equation (7.0.1) correspond to the problems of determination of the string characteristics (its length and density) with such alternating density.

Chapter 8.

Inverse problems connected with determination of arbitrary set of point sources

This chapter is devoted to the following problem set in Megrabov (1977b). Suppose in the half-space with the free boundary or in the whole space x, y, z filled with the medium, the N sources (the oscillators, or the emitters) are arbitrarily disposed which generate the summary vibration field u in the medium satisfying the wave equation. Each source has own impulse form and may be switched on and switched off (repeatedly) in arbitrary moments of time. The number of sources N, their coordinates and impulse forms are unknown. We need to find whether such point set E exists in the boundary of the half-space (in the case of the space — on the arbitrary plane such that all the sources are located from the one side of this plane) so that the field u set in E allows us to solve the inverse problem uniquely. The inverse problem in the nonstationary case is to find the number N, the coordinates and form of the impulse of each source. In the stationary case we must find the number N, the coordinates, the amplitude, phase and frequency of each source.

We consider also here the static inverse problem on determination of the number of point sources of Newton potential, their coordinates and masses. In this case, we suppose that the masses may have both signs.

In this chapter we show that in all of these three variants: nonstationary, stationary and static, such set E exists. We give also the examples of such sets. The two cases in all these variants are considered. In the first case the number N is unknown but we know that $N \leq n$, where n is known. This is the case 1. In the case 2 N and n are unknown.

In the case 2, the set E may be, in particular, a plane domain, extremely small, or some countable set of points. The field u set in a countable set of points is apparently, a minimal information provided a unique solution of the inverse problem since the number N of point sources is known and may be extremely large.

In the case 1 we show that the set E is the set consisted of a finite number of points dependent on n. In particular, as the set E, we can take the following set consisted of $\{(C_{2n}^2 + 1)f(2n)\}$ points. We take in the plane mentioned above $(C_{2n}^2 + 1)$ arbitrary straight lines. In each of these lines we take $f(2n)$ arbitrary points. The number C_n^m is the binomial coefficient (the number of combinations from n elements on m), and $f(n) = n[n(n^2 - 1)/2 + 1]$. In certain cases the number of points is the set E can be diminished to $3f(2n)$; $2f(2n)$ and even $f(2n)$.

In the case 1 the research of the inverse problems is based on applying certain results from the theory of Tchebysheff system of functions (T-systems). In the case 2 we use the analytical extension of the field u with respect to one of spatial variables.

The inverse problems considered in this chapter are the problems when we seek the right-hand side of special form responded to an arbitrary number of point emissive sources or point masses in the wave equation or in the Poisson equation. These problems are close to the inverse potential problems (Prilepko, 1973) and to the works of Alekseev and Megrabov (1977), Isakov (1977), Zapreev (1976), Kardakov (1976), Bleistein and Cohen (1977).

In § 8.12 the possible variants of application of the researches are discussed. These applications are in geophysics, acoustics, bioacoustics and those areas of biophysics where the sources of radiation are the living organisms.

8.1. DIRECT PROBLEM AND ITS SOLUTION

Here and further in this chapter N and n are arbitrary natural numbers, $N \leq n$; $\mu = \{M_j\}$ is the set consisted of N points M_j with the Cartesian coordinates x_j, y_j, z_j arbitrarily disposed in the half-space $z > 0$ (i. e. $z_j > 0$); $\varphi_1(t), \varphi_2(t), \ldots, \varphi_N(t)$ is a finite sequence of real functions defined for $-\infty < t < \infty$ satisfying the condition $\varphi_j(t) \not\equiv 0$; t_j, m_j are real numbers; $j = 1, 2, \ldots, N$; v is a positive number. Suppose also that all M_j are distinct.

Direct problem 8.1 (nonstationary case).

In the half-space $z > 0$ (or in the whole space) we consider the equation

$$\frac{\partial^2 u}{\partial x^2} + \frac{\partial^2 u}{\partial y^2} + \frac{\partial^2 u}{\partial z^2} = \frac{1}{v^2}\frac{\partial^2 u}{\partial t^2} - \sum_{j=1}^{N} \varphi_j(t)\delta(M - M_j), \tag{8.1.1}$$

where M is the point the coordinates x, y, z; δ is the three-dimensional Dirac function;

$$\varphi_j(t)|_{t<t_j} = 0,$$

and the initial conditions are as follows

$$u\Big|_{t<T} = 0, \quad u\Big|_{t=T} = \frac{\partial u}{\partial t}\Big|_{t=T} = 0, \quad T = \min_{1\le j\le N} t_j. \tag{8.1.2}$$

Besides, in the case of the half-space we consider the boundary condition

$$\frac{\partial u}{\partial z}\Big|_{z=0} = 0. \tag{8.1.3}$$

Conditions (8.1.1)–(8.1.3) are set for $-\infty < x, y, t < \infty$. To find the function $u(x, y, z, t)$ which has continuous derivatives up to the second order everywhere except for the points $M_j \in \mu$ and satisfying conditions (8.1.1)–(8.1.3). The numbers v, N and the coordinates of the points M_j and the functions $\varphi_j(t)$ are set.

This problem can be considered as the problem of determining the vibration field excited by the family of N point sources (emitters). The source in the point M_j has the form of impulse $\varphi_j(t)$ and is turning on the moment t_j. The sources are numerated in a certain order. The time t is counted from the moment of beginning the vibrations in the plane $z = 0$ (we can take another moment as $t = 0$) (see Figure 8.1a).

Note that the number of sources radiating simultaneously can vary during the process since each source can turn on and switch off in an arbitrary moment of time and more than once. This phenomena is set by the functions $\varphi_j(t)$. The number N is the summary number of sources which switch on the time interval $T \le t < \infty$, of course, is independent of the time t. This will hold both for direct and inverse problems; the functions $\varphi_j(t)$ must satisfy only the smoothness conditions.

Suppose each of the functions $\varphi_j(t)$ has continuous derivatives up to the second order for $-\infty < t < \infty$. As it follows from the known formula for

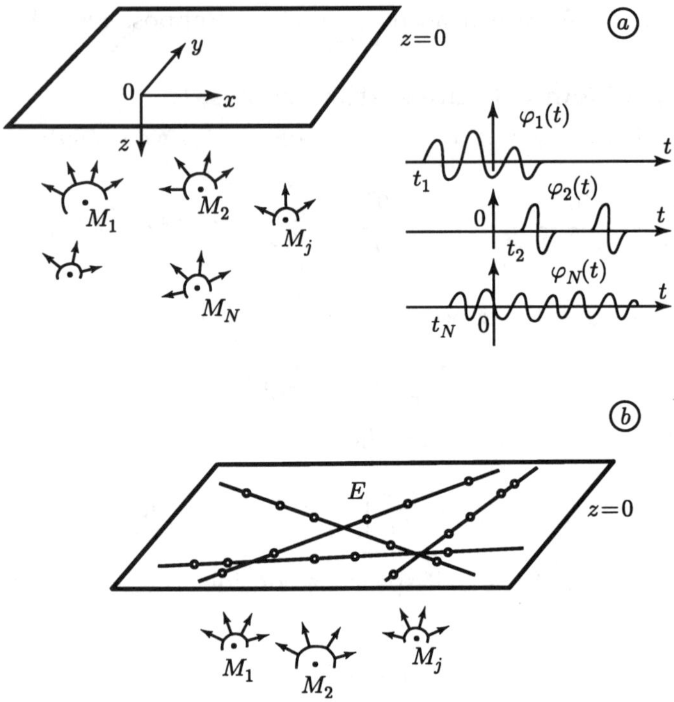

Figure 8.1. a) To the direct problem with an arbitrary set of point sources (emitters). b) The example of the set E where the field u is set in the general case

the field of the point source (Koshlyakov, et al., 1970, p. 108), the direct problem 8.1 has the following solution in the case of the half-space

$$u(x,y,z,t) = \frac{1}{4\pi}\sum_{j=1}^{N}\frac{\varphi_j\{t - R_j(x,y,z)/v\}}{R_j(x,y,z)} + \psi(x,y,z,t), \tag{8.1.4}$$

where

$$\psi(x,y,z,t) = \frac{1}{4\pi}\sum_{j=1}^{N}\frac{\varphi_j\{t - R_j(x,y,-z)/v\}}{R_j(x,y,-z)},$$

and

$$R_j(x,y,z) = \{(x-x_j)^2 + (y-y_j)^2 + (z-z_j)^2\}^{1/2}$$

is the distance between the points M and M_j. The direct problem 8.1 for the case of the whole space has the following solution

$$u(x,y,z,t) = \frac{1}{4\pi}\sum_{j=1}^{N}\frac{\varphi_j\{t - R_j(x,y,z)/v\}}{R_j(x,y,z)}. \tag{8.1.5}$$

The first term in (8.1.4) and the function (8.1.5) correspond to the incident (radiated, primary) wave and is represented by the collection of spherical waves with the form $\varphi_j(t)$ extending from the points M_j. The function ψ is the reflected wave from the boundary $z = 0$. In both cases, we have

$$u(x,y,0,t) = \frac{1}{2\pi l}\sum_{j=1}^{N}\frac{\varphi_j\{t - R_{0j}(x,y)/v\}}{R_{0j}(x,y)}, \tag{8.1.6}$$

where $l = 1$ in the case of the half-space, and $l = 2$ in the case of the space. The distance between the point M from the plane $z = 0$ and the point M_j is denoted by

$$R_{0j}(x,y) = \{(x-x_j)^2 + (y-y_j)^2 + z_j^2\}^{1/2}.$$

Suppose that the function $\varphi_j(t)$ is absolutely integrable in the line $-\infty < t < \infty$ for $j = 1, 2, \dots, N$. Then

$$u(x,y,0,t) = \frac{1}{2\pi l}\int_{-\infty}^{\infty} U(x,y,k)e^{ikt}\,dk, \tag{8.1.7}$$

where

$$U(x,y,k) = \sum_{j=1}^{N}\Phi_j(k)\,\frac{e^{-i(k/v)R_{0j}(x,y)}}{R_{0j}(x,y)}. \tag{8.1.8}$$

Here $\Phi_j(k)$ is the Fourier transform of the function $\varphi_j(t)$, and the inversion formula holds

$$U(x,y,k) = l\int_{-\infty}^{\infty} u(x,y,0,t)e^{-ikt}\,dt. \tag{8.1.9}$$

The stationary case. Let $N, v, \mu = \{M_j\}$, $R_j(x,y,z)$ be defined as in § 8.1; k_j are arbitrary numbers (may be complex); Φ_j are arbitrary complex numbers ($\Phi_j \neq 0$): $\Phi_j = A_j e^{i\theta_j}$, $0 < A_j < \infty$, $0 \leq \theta_j < 2\pi$ ($j = 1, 2, \dots, N$). The functions

$$u(x,y,z,t) = \frac{1}{4\pi}\sum_{j=1}^{N}\Phi_j\,\frac{e^{-i(k_j/v)R_j(x,y,z)}}{R_j(x,y,z)}\,e^{ik_jt} + \frac{1}{4\pi}\sum_{j=1}^{N}\Phi_j\,\frac{e^{-i(k_j/v)R_j(x,y,-z)}}{R_j(x,y,-z)}\,e^{ik_jt}, \tag{8.1.10}$$

$$u(x,y,z,t) = \frac{1}{4\pi} \sum_{j=1}^{N} \Phi_j(k) \frac{e^{-i(k_j/v)R_j(x,y,z)}}{R_j(x,y,z)} e^{ik_j t} \tag{8.1.11}$$

are the solutions of the stationary analog of the direct problem 8.1 in the cases of the half-space and the whole space respectively. Here each (stationary) source situated in the point M_j has the form of impulse $A_j e^{i(k_j t+\theta_j)}$ with the frequency k_j, the amplitude A_j and the initial phase θ_j. The time t is counted from an arbitrary moment. In both cases we have

$$u(x,y,0,t) = \frac{1}{2\pi l} \sum_{j=1}^{N} \Phi_j \, e^{ik_j t} \frac{e^{-i(k_j/v)R_{0j}(x,y)}}{R_{0j}(x,y)}, \tag{8.1.12}$$

where $l = 1$ in the case of the half-space and $l = 0$ in the case of the whole space.

The static case. Consider also the function

$$u(x,y,z) = \sum_{j=1}^{N} m_j \Big\{ \frac{1}{R_j(x,y,z)} + \frac{\rho}{R_j(x,y,-z)} \Big\}, \tag{8.1.13}$$

where $\rho = 0$ or $\rho = 1$. In both cases

$$u(x,y,0) = l \sum_{j=1}^{N} \frac{m_j}{R_{0j}(x,y)}, \tag{8.1.14}$$

where $l = 1 + \rho$. In the case $\rho = 0$, the function $u(x,y,z)$ is the summary Newton potential (to within the constant multiplier) of the set of point masses m_j concentrated in the points M_j. Note, however, that in our research the masses m_j in formulas (8.1.13), (8.1.14) may be both positive and negative.

8.2. SOME AUXILIARY GEOMETRICAL DEFINITIONS

The constructions give below are similar for nonstationary, stationary and static problems and for both cases 1 and 2.

We denote the plane $z = 0$ by S. Suppose $\Pi \subset S$ is a certain straight line. Through each point $M_j \in \mu$ we carry on the plane normal to the line Π and construct the semicircle $O_{j\Pi}$ in this plane with the center in the line Π

passing through the point M_j and lying in the half-space $z \geq 0$. The set of all semicircles $O_{j\Pi}$ we denote by $O_{\Pi,\mu}$. Set

$$a_j(t) = \varphi_{j_1}(t) + \varphi_{j_2}(t) + \ldots + \varphi_{j_m}(t),$$
$$\Psi_j = \Phi_{j_1} e^{ik_{j_1}t} + \Phi_{j_2} e^{ik_{j_2}t} + \ldots + \Phi_{j_m} e^{ik_{j_m}t}, \tag{8.2.1}$$
$$c_j = m_{j_1} + m_{j_2} + \ldots + m_{j_m},$$

where $j_1, j_2, \ldots, j_m$, $1 \leq m \leq N$ are the numbers of those points $M_j \in \mu$ which lie in one semicircle $O_{j\Pi}$ from the set $O_{\Pi,\mu}$. Some semicircles from $O_{\Pi,\mu}$ may coincide if there exist the points M_j equidistant from the line Π and lying in one plane normal to Π. In each such semicircle there lie several points M_j.

Let Π_1 and Π_2 be noncoinciding lines in the plane $z = 0$. Construct for each of them the sets $O_{\Pi_1,\mu}$ and $O_{\Pi_2,\mu}$. The set of all points of intersection of these two sets we denote by $\mu_{1,2}$:

$$\mu_{1,2} = O_{\Pi_1,\mu} \cap O_{\Pi_2,\mu}.$$

Evidently, the set $\mu_{1,2}$ is finite and contains no more than N^2 points. Since, through each point $M_j \in \mu$, at least one semicircle from each set is passing, the set $\mu_{1,2}$ is nonempty and contains μ: $\mu \subset \mu_{1,2}$. Note that the set $\mu_{1,2}$, generally speaking, is wider than μ. This means that the set $\mu_{1,2}$ does not coincide with μ since the sets $O_{\Pi_1,\mu}$ and $O_{\Pi_2,\mu}$ may intersect in the points which does not belong to μ; however, for arbitrary choice Π_1 and Π_2 this is improvable.

We shall say that the pair of noncoinciding lines Π_1 and Π_2 has the property B_μ (relatively the set μ), if in each pair of intersecting semicircles contained the semicircles from each family $O_{\Pi_1,\mu}$, $O_{\Pi_2,\mu}$ there exists at least one semicircle from a family which does not intersect with the semicircles of the other family. Evidently, if the pair of lines Π_1 and Π_2 satisfies the property B_μ, then the sets $\mu_{1,2}$ and μ coincides: $\mu_{1,2} = \mu$ (each point from $\mu_{1,2}$ will be the point from μ since each of the semicircles contains at most one point from μ).

If the set μ is unknown and the lines Π_1 and Π_2 are given, then the property B_μ may fail. However, if μ is unknown, and we find the sets $O_{\Pi_1,\mu}$ and $O_{\Pi_2,\mu}$, then we may say whether the lines Π_1 and Π_2 have the property B_μ.

Further, we find the line of intersection of the plane S with the plane normal to the segment connected the arbitrary pair of point of the set μ and dividing it in two. The set Π_μ of all such lines contains no more than C_N^2 elements. The following statement holds.

Statement 8.2.1. In order that for at most in one semicircle $O_{j\Pi} \in O_{\Pi,\mu}$ lie at most two points from the set μ it is necessary and sufficient that $\Pi \subset \Pi_\mu$.

From this statement the following corollary follows.

Corollary 8.2.1. *If* $\Pi \not\subset \Pi_\mu$ *then in each semicircle* $O_{j\Pi}$ *the only one point* $M_j \in \mu$ *lies and* $a_j(t) = \varphi_j(t)$, $\Psi_j = \Phi_j$, $c_j = m_j$.

8.3. THE AUXILIARY RESULTS FOR THE CASE 1 CONNECTED WITH THE T-SYSTEMS

8.3.1. ***T*-systems and their properties.** We shall formulate here some results from the theory of Tchebysheff T-systems. More detailed account on these systems may be found in Remez (1969), Karlin and Studden (1973).

We say that n real continuous functions $\psi_1(x), \psi_2(x), \dots, \psi_n(x)$ linearly independent in the segment $[a, b]$ form Tchebysheff system of functions (or, briefly, the T-system) of order $(n-1)$ in the segment $[a, b]$, if each generalized polynomial

$$F_n(x) = \sum_{k=1}^{n} a_k \psi_k(x) \qquad \Big(\sum_{k=1}^{n} |a_k| > 0\Big)$$

may have no more than $(n-1)$ zeros in the segment $[a, b]$.

The special case of T-systems is the case of TM-systems (Tchebysheff—Markov).

If we have n functions

$$\psi_1(x),\ \psi_2(x),\ \dots,\ \psi_n(x), \tag{8.3.1}$$

then they form the TM-system of the order $(n-1)$ in the segment $[a, b]$, if for each value $\nu = 1, 2, \dots, n$, the functions

$$\psi_1(x),\ \psi_2(x),\ \dots,\ \psi_\nu(x),$$

form the T-system of order $(\nu - 1)$ in the segment $[a, b]$. The following statement holds.

Theorem on *TM*-systems. *Each system of functions* (8.3.1) *which are* n *times continuously differentiable in* $[a, b]$ *and satisfying the* W*-condition is a* TM*-system of* $(n-1)$ *order in the segment* $[a, b]$.

We say that the system $\psi_1(x), \dots, \psi_n(x)$ satisfies the W-condition if each determinant

$$W(\psi_1) \equiv \psi_1; \qquad W(\psi_1, \psi_2) = \begin{vmatrix} \psi_1 & \psi_2 \\ \psi_1' & \psi_2' \end{vmatrix}; \quad \dots \quad ; \quad W(\psi_1, \psi_2, \dots, \psi_n),$$

where

$$W(\psi_1, \psi_2, \dots, \psi_k) = \begin{vmatrix} \psi_1 & \psi_2 & \dots & \psi_k \\ \psi_1' & \psi_2' & \dots & \psi_k' \\ \dots & \dots & \dots & \dots \\ \psi_1^{(k-1)} & \psi_2^{(k-1)} & \dots & \psi_k^{(k-1)} \end{vmatrix}, \qquad k = 1, 2, \dots, n$$

does not vanish in the segment $[a, b]$.

Here the prime denotes the differentiation with respect to x.

The proof of this theorem can be found in Karlin and Studden (1973, Chapter 11). The system of functions satisfying the W-condition were considered by Polya (1922). In this work he had investigated the integrals of correspondent linear differential equations whose fundamental system is system (8.3.1).

8.3.2. The number of zeros of Wronskian for our system of functions. Consider the system of N functions

$$\psi_k = \{(x - x_k)^2 + r_k^2\}^{-1/2} = w_k(x)^{-1/2} \ (> 0), \quad w_k(x) = (x - x_k)^2 + r_k^2,$$
$$x_k \in (-\infty, \infty), \qquad r_k > 0, \qquad k = 1, 2, \dots, N, \tag{8.3.2}$$

where all the pairs (x_k, r_k) are distinct. We can show that the system of functions $\{\psi_k(x)\}_1^N$ is linearly independent in each interval (a, b) from the x–axis. The following lemma holds.

Lemma 8.3.1. *Suppose all the pairs (x_k, r_k) are distinct. Then for each N natural, the Wronskian*

$$W(\psi_1, \psi_2, \dots, \psi_N) \equiv W_N(x) \equiv \begin{vmatrix} \psi_1 & \psi_2 & \dots & \psi_N \\ \psi_1' & \psi_2' & \dots & \psi_N' \\ \dots & \dots & \dots & \dots \\ \psi_1^{(N-1)} & \psi_2^{(N-1)} & \dots & \psi_N^{(N-1)} \end{vmatrix} \tag{8.3.3}$$

has in the real axis x (and also in the whole complex plane $z = x + iy$) no more than δ_N finite zeros, where

$$\delta_N = 3N(N-1)/2.$$

Proof. First we show that for each $k = 1, 2, \dots, N$ and $n = 0, 1, 2, \dots$ the formula

$$\psi_k^{(n)}(x) = w_k^{-1/2} P_{n,k}(x)/w_k^n \tag{8.3.4}$$

holds, where $P_{n,k}(x)$ is a certain polynomial of the degree n. To this end use the induction method.

For $n = 0, 1, 2$ formula (8.3.4) is established by immediate computation

$$\begin{array}{lll} (n=0) & \psi_k(x) = w_k^{-1/2} P_{0,k}/w_k^0, & P_{0,k} \equiv 1; \\ (n=1) & \psi_k'(x) = w_k^{-1/2} P_{1,k}(x)/w_k, & P_{1,k}(x) = -(x - x_k); \\ (n=2) & \psi_k''(x) = w_k^{-1/2} P_{2,k}(x)/w_k^2, & P_{2,k} = -w_k + 3(x - x_k)^2. \end{array}$$

Setting that formula (8.3.4) holds for n, prove it for $(n+1)$. So, suppose that

$$\psi_k^{(n)}(x) = w_k^{-1/2} P_{n,k}(x) w_k^{-n}.$$

Then, differentiating this formula, we obtain

$$\psi_k^{(n+1)}(x) = w_k^{-1/2} w_k^{-(n+1)} \{-(2n+1)(x - x_k) P_{n,k}(x) + w_k P_{n,k}'(x)\}.$$

The expression in the figure brackets is the polynomial $P_{n+1,k}(x)$ of the degree $(n+1)$; therefore

$$\begin{aligned} P_{n,k}(x) &= a_{n,k} x^n + \dots, \qquad a_{n,k} \neq 0, \\ P_{n,k}'(x) &= n a_{n,k} x^{n-1} + \dots. \end{aligned}$$

The coefficient before x^{n+1} in the figure brackets is equal to

$$-(2n+1) a_{n,k} + n a_{n,k} = -(n+1) a_{n,k} \neq 0.$$

Hence it follows that

$$\psi_k^{(n+1)}(x) = w_k^{-1/2} P_{n+1,k}(x)/w_k^{n+1}$$

whence formula (8.3.4) follows.

Write this formula in the form

$$\psi_k^{(n)} = w_k^{1/2-N} P_{n,k} w_k^{N-1-n}. \tag{8.3.5}$$

Substitute this expression into (8.3.3). Then we obtain

$$W_N(x) = \begin{vmatrix} a_{11} & a_{12} & \dots & a_{1N} \\ a_{21} & a_{22} & \dots & a_{2N} \\ \dots & \dots & \dots & \dots \\ a_{N1} & a_{N2} & \dots & a_{NN} \end{vmatrix}, \qquad a_{ij} = \psi_j^{(i-1)}.$$

In other words

$$W_N(x) = \prod_{k=1}^{N} w_k^{1/2-N} D_N(x),$$

where

$$D_N(x) \equiv \begin{vmatrix} w_1^{N-1} & w_2^{N-1} & \dots & w_N^{N-1} \\ P_{1,1}w_1^{N-2} & P_{1,2}w_2^{N-2} & \dots & P_{1,N}w_N^{N-2} \\ P_{2,1}w_1^{N-3} & P_{2,2}w_2^{N-3} & \dots & P_{2,N}w_N^{N-3} \\ \dots & \dots & \dots & \dots \\ P_{N-1,1}w_1^{N-N} & P_{N-1,2}w_2^{N-N} & \dots & P_{N-1,N}w_N^{N-N} \end{vmatrix}.$$

The element d_{ij} of the determinant $D_N(x)$ has the form

$$d_{ij} = P_{i-1,j}w_j^{N-i}, \qquad i,j = 1,2,\dots,N. \tag{8.3.6}$$

Each term of the determinant D_N is the product

$$d_{\alpha_1 1}d_{\alpha_2 2}\cdots d_{\alpha_N N}$$

to within the multiplier (± 1). Here $\alpha_1, \alpha_2, \dots, \alpha_N$ is a certain transposition of the numbers $1, 2, \dots, N$. Taking into account formula (8.3.6) we obtain that each term of the determinant D_N is the product

$$\{P_{\alpha_1-1,1}P_{\alpha_2-1,2}\cdots P_{\alpha_N-1,N}\}\,\{w_1^{N-\alpha_1}w_2^{N-\alpha_2}\cdots w_N^{N-\alpha_N}\}.$$

The first product of this expression is the polynomial of the degree

$$\alpha_1 - 1 + \alpha_2 - 1 + \ldots + \alpha_N - 1 = \alpha_1 + \alpha_2 + \ldots + \alpha_N - N = N(N-1)/2$$

since

$$\alpha_1 + \alpha_2 + \ldots + \alpha_N = 1 + 2 + \ldots + N = N(N+1)/2.$$

The second product is the polynomial of the degree

$$2\{N-\alpha_1+N-\alpha_2+\ldots+N-\alpha_N\} = 2\{N^2-(\alpha_1+\alpha_2+\ldots+\alpha_N)\} = N(N-1).$$

Therefore, each term of the determinant D_N is a polynomial of the degree

$$\delta_N = 3N(N-1)/2.$$

Hence it follows that D_N is a polynomial of the degree no more than δ_N. Since the functions of the system $\{\varphi_k(x)\}_1^N$ are linearly independent in each (arbitrarily small) interval in the x-axis, then $W_N(x) \not\equiv 0$[i)], therefore, $D_N(x) \not\equiv 0$. From here it follows that not all the coefficients of

$$D_N(x) = a_1 x^{\delta_N} + a_2 x^{\delta_N - 1} + \ldots + a_{\delta_N + 1}$$

are equal to zero.

Taking into account the fundamental algebraic theorem, the polynomial $D_N(x)$ (and, therefore, the determinant $W_N(x)$) has in the complex plane (and, therefore, in the real axis x) no more than δ_N finite zeros. The lemma is proved. □

8.3.3. The number of zeros of the polynomial $F_n(x) = \sum_{k=1}^{n} a_k \psi_k(x)$.

Lemma 8.3.2. *(The basic lemma). Consider the function*

$$F_n(x) = \sum_{k=1}^{n} a_k \psi_k(x), \quad \psi_k(x) = \{(x - x_k)^2 + r_k^2\}^{-1/2} > 0, \tag{8.3.7}$$

where a_k, x_k, r_k are real numbers, so that

$$a_k \in (-\infty, \infty), \quad x_k \in (-\infty, \infty), \quad r_k > 0.$$

Suppose also that all the pairs (x_k, r_k), $k = 1, 2, \ldots, n$ are distinct, and not all a_k $(k = 1, 2, \ldots, n)$ are equal to zero.

Then, for each n natural, the generalized polynomial $F_n(x)$ of form (8.3.7) may have in the real axis x no more than $f(n) - 1$ finite zeros, where

$$f(n) = (\Delta_n + 1)n, \qquad \Delta_n = \sum_{k=1}^{n} \delta_k = n(n^2 - 1)/2, \tag{8.3.8}$$

so that

$$f(n) = n\{n(n^2 - 1)/2 + 1\}.$$

[i)] It is the corollary of the following general theorem (perhaps, it is known). Let the functions $y_1(x), y_2(x), \ldots, y_n(x)$ be linearly independent in each subinterval of interval $[a, b]$ and derivatives $y_1^{(k)}, y_2^{(k)}, \ldots, y_m^{(k)}$ are continuous in $[a, b]$. Then each Wronskian $W_k(x) = W(y_1, y_2, \ldots, y_k)$, $k = 1, 2, \ldots, n$, is not identically equal to zero in $[a, b]$.

Proof. Assume the contrary, i. e., $F_n(x)$ vanishes at most in $f(n)$ finite points of the real axis. Denote this point set by X_n. Since $f(n) = (\Delta_n + 1)n$, then, evidently, there exists such a set of finite segments J_s from the x-axis ($s = 1, 2, \dots, \Delta_n + 1$) (these segments are not intersecting) that each segment J_s contains n points exactly from the set X_n. By Lemma 8.3.1, the complete number of zeros of all the Wronskian $W_1, W_2, \dots, W_n$ from the W-condition is no more than Δ_n defined in formula (8.3.8). Therefore, for at most one of $(\Delta_n + 1)$ segments J_s there are no zeros of the determinants W_k, $k = 1, 2, \dots, n$; hence the W-condition holds. Taking into account the theorem on TM-systems (see p. 3.1), in such segment the functions $\psi_1(x), \psi_2(x), \dots, \psi_n(x)$ form the TM-system of the order $(n-1)$. Therefore, the polynomial $F_n(x)$ has in this segment no more than $(n-1)$ zeros. On the other hand, this segment J_s contains n points of the set X_n, i. e. n zeros of the function $F_n(x)$. This contradiction proves the lemma. □

Corollary 8.3.1. *If a certain polynomial $F_n(x)$ of form* (8.3.7), *where n is natural, has in the whole axis x not less than $f(n)$ finite zeros, then $a_k = 0$ for $k = 1, 2, \dots, n$ and $F_n(x) \equiv 0$.*

8.3.4. Ordering of the set of pairs $\{x_k, r_k\}$ by increasing. Consider the set of pairs (x_k, r_k) where $x_k \in (-\infty, \infty)$, $r_k > 0$, $k = 1, 2, \dots, N$. Suppose all these pairs are distinct, i. e. for each two pairs (x_{k_1}, r_{k_1}) and (x_{k_2}, r_{k_2}) at least one of the inequalities

$$x_{k_1} \neq x_{k_2}, \qquad r_{k_1} \neq r_{k_2}$$

holds.

We shall say that the set of pairs $\{x_k, r_k\}_1^N$, is ordered by increasing if the pairs (x_k, r_k) are numerated by nondecreasing x_k, i. e. $x_1 \leq x_2 \leq \dots \leq x_N$, where, for equal x_k (if there are such), r_k is increasing. Evidently, each finite set $\{x_k, r_k\}_1^N$ can be ordered in such a way. The following lemma holds.

Lemma 8.3.3. *Suppose each of two sets of pairs $\{x_k, r_k\}_1^N$ and $\{x'_k, r'_k\}_1^N$ has no equal pairs (inside itself) and is ordered by increasing. Then the equalities*

$$x_j = x'_{k_j}, \quad r_j = r'_{k_j}, \quad \text{where } j = 1, 2, \dots, N \tag{8.3.9}$$

hold if and only if

$$j = k_j, \quad j = 1, 2, \dots, N.$$

Here $\{k_j\}_1^N$ is a certain transposition of the numbers $1, 2, \dots, N$.

Proof. Assume the contrary. This means that there exists such $j = 1, 2, \dots, N$ that

$$x_j = x'_{k_j}, \quad r_j = r'_{k_j}, \quad \text{where} \quad k_j \neq j. \tag{8.3.10}$$

a) Let, for example, $k_j > j$. Then, according to (8.3.9) for each pair from the set $\{x'_k, r'_k\}_1^N$, there exists the equal pair from the set $\{x_k, r_k\}_1^N$. Therefore, there exists at most one pair (x'_p, r'_p), $(p < k_j)$ from the set (x'_k, r'_k), $k = 1, 2, \dots, k_j - 1$, so that

$$x_s = x'_p, \quad r_s = r'_p, \quad \text{where} \quad s \geq j. \tag{8.3.11}$$

We have $s \neq j$, since, otherwise, by (8.3.10) and (8.3.11), the pairs (x'_p, r'_p) and (x'_{k_j}, r'_{k_j}) will coincide, which is impossible (we consider the sets which contains distinct pairs). Therefore, we should consider only the case $s > j$. In this case, one of the conditions must hold

$$x_j < x_s; \quad x_j = x_s, \quad r_j < r_s.$$

Let, for example, $x_j < x_s$. Then we have

$$x'_p = x_s > x_j = x'_{k_j}, \quad \text{i.e.} \quad x'_p > x'_{k_j},$$

which is impossible since $p < k_j$.

The analogous contradiction holds in the case when $x_j = x_s$, $r_j < r_s$. In this case, we have

$$x'_p = x_s = x_j = x'_{k_j}$$

so that all x'_k from $k = p$ up to $k = k_j$ are equal to x'_p. But then we must have $r'_p < r'_{k_j}$. From the supposition $r_j < r_s$ we obtain the opposite inequality

$$r'_p = r_s > r_j = r'_{k_j}.$$

So we had obtained the contradictions in both possible cases.

b) We consider now the case when $k_j < j$. Then there exist such pairs (x_p, r_p) and (x'_s, r_s) that

$$x_p = x'_s, \quad r_p = r'_s, \quad \text{where} \quad p < j, \quad s > k_j.$$

Therefore, at most one of the conditions

$$x_p < x_j; \quad x_p = x_j, \quad r_p < r_j$$

holds. Suppose, at first, that $x_p < x_j$. Then

$$x'_s = x_p < x_j = x'_{k_j}.$$

This means that $x'_s < x'_{k_j}$, where $s > k_j$, which is impossible.

If we set that $x_p = x_j$, $r_p < r_j$, then all x'_k, where k is changing from k_j up to s, are equal, and $r_s > r_{k_j}$. However, from the supposition, we obtain

$$r_s = r_p < r_j = r_{k_j}.$$

Again we obtain the contradiction. The lemma is proved. □

8.3.5. The uniqueness theorem for generalized polynomials. Evidently, each polynomial, not equal to zero identically

$$F_N(x) = \sum_{k=1}^{N} a_k \psi_k(x), \quad \psi_k(x) = \{(x - x_k)^2 + r_k^2\}^{-1/2} > 0$$

(the pairs (x_k, r_k) are distinct) can be represented as the sum $\sum_{k=1}^{N_1 \le N} a_k \psi_k(x)$, where $a_k \neq 0$ for $k = 1, 2, \ldots, N_1$, and the pairs of the set $\{x_k, r_k\}_1^{N_1}$ are ordered by decreasing.

The following theorem holds.

Theorem 8.3.1. *Suppose N_1, N'_1, n are arbitrary natural numbers so that $N_1 \le n$, $N'_1 \le n$, and*

$$F_{N_1}(x) = \sum_{k=1}^{N_1 \le n} a_k \psi_k(x), \quad \psi_k(x) = \frac{1}{\sqrt{(x - x_k)^2 + r_k^2}}, \tag{8.3.12}$$

$$F_{N'_1}(x) = \sum_{k=1}^{N'_1 \le n} a'_k \psi'_k(x), \quad \psi'_k(x) = \frac{1}{\sqrt{(x - x'_k)^2 + (r'_k)^2}}, \tag{8.3.13}$$

are two generalized polynomials. Without loss of generality, we suppose that each of the sets of pairs $\{x_k, r_k\}_1^{N_1}$ and $\{x'_k, r'_k\}_1^{N'_1}$ has not equal pairs inside itself; the pairs are ordered by decreasing; and

$$\begin{aligned} &a_k \neq 0, \;\; x_k \in (-\infty, \infty), \;\; r_k > 0, \quad \text{for} \quad k = 1, 2, \ldots, N_1; \\ &a'_k \neq 0, \;\; x'_k \in (-\infty, \infty), \;\; r'_k > 0, \quad \text{for} \quad k = 1, 2, \ldots, N'_1; \end{aligned}$$

If in $f(2n)$ arbitrary points of the real axis $F_{N_1}(x) = F_{N'_1}(x)$, then, $N_1 = N'_1$, $a_k = a'_k$, $x_k = x'_k$, $r_k = r'_k$ for $k = 1, 2, \ldots, N_1 = N'_1$; therefore, $F_{N_1(x)} \equiv F_{N'_1}(x)$ in the whole x-axis. Here we have

$$f(n) = n\{n(n^2 - 1)/2 + 1\}. \tag{8.3.14}$$

Proof. Evidently, the difference $F_{N_1} - F_{N_1'}(x)$ is the generalized polynomial of the order no more than $2n$ (so that the number of distinct pairs (x_k, r_k), (x_k', r_k') generating the set of linearly independent functions $\psi_k(x)$, $\psi_k'(x)$ of form (8.3.12), (8.3.13) is no greater than $2n$). Denote this difference, therefore, by $F_{2n}(x)$. By the condition of Theorem 8.3.1, the function $F_{2n}(x)$ has more than $\{f(2n) - 1\}$ zeros; therefore, by Corollary 8.3.1, after combining the terms in $F_{2n}(x)$ with equal pairs (x_k, r_k) and (x_k', r_k') (if there are such pairs) we shall obtain that all its coefficients b_j are equal to zero. Hence it follows that $F_{2n}(x) \equiv 0$, i. e. $F_{N_1}(z) \equiv F_{N_1'}(x)$ in the whole x-axis.

Evidently, the function F_{2n} can be represented as follows

$$F_{2n}(x) = \sum_{j=1}^{N''\le N_1} (a_{k_j} - a'_{k'_j})\psi_{k_j} + \sum_{j=N''+1}^{N_1} a_{k_j}\psi_{k_j} + \sum_{j=N''+1}^{N_1'} a'_{k'_j}\psi'_{k'_j}. \quad (8.3.15)$$

Here the first sum (with the coefficients $b_j = a_{k_j} - a'_{k'_j}$) contains the differences of only those terms of polynomials $F_{N_1}(x)$ and $F_{N_1'}(x)$ (their number $N'' \le N_1$) which satisfy the condition

$$x_{k_j} = x'_{k'_j}, \quad r_{k_j} = r'_{k'_j} \quad (j = 1, 2, \dots, N'' \le N_1).$$

Here the sequence $k_1, k_2, \dots, k_{N''}$ increases, and its terms belong to the sequence $1, 2, \dots, N_1'$. The sequence $k_1', k_2', \dots, k'_{N_1'}$ is a transposition of the numbers $1, 2, \dots, N_1'$. However, in two last sums in (8.3.15) we must have $N_1 = N'' = N_1'$, so these sums are equal to zero. Really, otherwise, since all the pairs (x_{k_j}, r_{k_j}) and $(x'_{k'_j}, r'_{k'_j})$ in these three sums in expression (8.3.15) are distinct (inside each of the sets of pairs $\{x_k, r_k\}_1^{N_1}$, $\{x_k', r_k'\}_1^{N_1'}$), then, at least one of the coefficient b_j of the polynomial $F_{2n}(x)$ coincides with a coefficient a_{k_j} or $a'_{k'_j}$ of the polynomial F_{N_1} or $F_{N_1'}$. But this is impossible, since $b_j = 0$, and, by the condition, $a_k \ne 0$, $a_k' \ne 0$ for all $k = 1, 2, \dots, N_1(N_1')$. Thus, $F_{2n}(x)$ is equal to the first sum in expression (8.3.15) for $N'' = N_1 = N_1'$ and $k_1, k_2, \dots, k_{N''}$ is the sequence $1, 2, \dots, N_1$; i. e., $k_j = j$. So, $F_{2n}(x)$ has the form

$$F_{2n}(x) = \sum_{j=1}^{N_1} (a_j - a'_{k'_j})\psi_k(x)$$

and $b_j = a_j - a'_{k'_j} = 0$, $x_j = x'_{k'_j}$, $r_j = r'_{k'_j}$ for $j = 1, 2, \dots, N_1$, where $k_1', k_2', \dots, k'_{N_1}$ is a certain transposition of the numbers $1, 2, \dots, N_1$. Since, each of the sets $\{x_k, r_k\}_1^{N_1}$ and $\{x'_k, r'_k\}_1^{N_1}$ has no equal pairs and is ordered by increasing, then, by Lemma 8.3.3, we have $k'_j = j$. Therefore,

$$a_j = a'_j, \quad x_j = x'_j, \quad r_j = r'_j, \quad j = 1, 2, \dots, N_1$$

which was to be proved. □

This theorem yields the following corollary.

Corollary 8.3.2. *Let $N \leq n$, where the natural n is given and N is unknown. Set a polynomial*

$$F_N(x) = \sum_{k=1}^{N} a_k \psi_k(x), \qquad \psi_k(x) = \frac{1}{\sqrt{(x - x_k)^2 + r_k^2}},$$

in $f(2n)$ arbitrary points of the x-axis ($f(n)$ is defined in (8.3.14)). Here $a_k \in (-\infty, \infty)$, $x_k \in (-\infty, \infty)$, $r_k > 0$; all the pairs (x_k, r_k) are distinct ($k = 1, 2, \dots, N$). Then we define uniquely all the triples (a_k, x_k, r_k), where $a_k \neq 0$, and the number N_1 ($\leq N$) of these triples.

8.4. PRELIMINARY REMARKS ON SOLUTIONS OF THE INVERSE PROBLEMS

Take in the plane $z = 0$ an arbitrary straight line Π and choose the Cartesian system (x, y) so that Π coincides with a certain line $y_0 =$ const. Suppose that x_j, y_j, z_j are coordinates of the point M_j in this coordinate system. Set

$$r_j = \{(y_0 - y_j)^2 + z_j^2\}^{1/2} > 0.$$

Suppose also that a pair $x_j, r_j > 0$ is given. The set of points from $z \geq 0$ which satisfy the equalities

$$x = x_j, \qquad \{(y_0 - y)^2 + z^2\} = r_j$$

is the semicircle $O_{j\Pi}$ (see § 2). This semicircle lives in the plane $x = x_j$ normal to the line Π with the center in Π and the radius r_j. Each of the semicircles $O_{j\Pi}$ is determined completely by the pair x_j, r_j.

Research of each the inverse problems considered below (§ 8.7, § 8.9, § 8.10) — static, nonstationary, and stationary is divided into two stages.

On the first stage, for each straight line Π_j from a certain set of straight lines $\{\Pi_j\}$ from the plane $z = 0$, we define the set of triples $\{D_j, x_j, r_j\}$, or the set of pairs $\{D_j,\ O_{j\Pi_i} \in O_{\Pi_{i\mu}}\}$ where $D_j \neq 0$. The number of these triples is equal to N_1 ($N_1 \leq N$) and is defined also. Here D_j is one of the values c_j or m_j (in the static case), $a_j(t)$ or $\varphi_j(t)$ (in the nonstationary case), and Ψ_j or Φ_j (in the stationary case). The values c_j, $a_j(t)$ and Ψ_j are linear combinations of the values m_j, $\varphi_j(t)$ and Φ_j respectively defined by equalities (8.2.1). In this case, the field u in each of the problems is set in a certain set of point from an arbitrary straight line $\Pi_i \in S$ (see § 2), which belongs to a certain set of straight lines. This stage has the distinctions for the static, nonstationary, stationary cases, and for the cases 1 and 2.

On the second stage, by the sets $\{D_j\}$, $O_{\Pi_{i\mu}}$, we recover all the desired values or the sets $\{D_j\}$, μ_{12}. The solution of the inverse problem on the second stage is similar for all the cases.

8.5. THE STATIC INVERSE PROBLEM WITH THE DATA ON THE STRAIT LINE

Suppose Π is an arbitrary straight line in the plane $z = 0$. Repeat the considerations of the first part of § 8.4. Evidently the field $u(x, y, 0)$ of form (8.1.14) in Π is the function dependent on x as follows:

$$u(x,y,0) = U_N(x) = \sum_{j=1}^{N} m_j \psi_j(x), \quad \psi_j(x) = 1/\sqrt{(x-x_j)^2 + r_j^2}. \tag{8.5.1}$$

In this sum, some pairs (x_j, r_j), generally speaking, may coincide. Among the pairs (x_j, r_j) are the equal pairs if and only if, for certain points M_j, the semicircles $O_{j\Pi}$ coincide, i. e., certain points M_j lie in one semicircle $O_{j\Pi}$ (see § 8.2).

Statement 8.2.1 yield the following statement.

Remark 8.5.1. Among the pairs (x_j, r_j) in equality (8.5.1) there are no equal pairs (and among the semicircles $O_{j\Pi}$ there are no coinciding semicircles) if and only if $\Pi \subseteq \Pi_\mu$.

Everywhere below we set that n is natural and such that $N \leq n$; $X_{n\Pi}$ is a set consisted from $f(2n)$ arbitrary distinct points from the plane $z = 0$ lying in the straight line Π. We have $f(n) = n\{n(n^2-1)/2 + 1\}$. We set also that

$$c_j = m_{j_1} + m_{j_2} + \ldots + m_{j_m}, \tag{8.5.2}$$

where $j_1, j_2, \ldots, j_m$ ($1 \leq m \leq N$) are the numbers of those points $M_j \in M$ which lie in one semicircle $O_{j\Pi} \in O_{\Pi,\mu}$.

The sum in equality (8.5.1) can be written in the form (if we combine the terms with equal pairs (x_j, r_j))

$$U_N(x) = \sum_{j=1}^{\tilde{N} \leq N} c_j \psi_j(x),$$

where all the pairs (x_j, r_j) are distinct. Therefore, Corollary 8.3.2 yields the following theorem.

Theorem 8.5.1. *Suppose the natural N is unknown and the coordinates of the points M_j and the numbers m_j, $j = 1, 2, \ldots, N$ are unknown also. The number n is known. Suppose we set the field $u(x, y, z)$ of form (8.1.13) in the set $X_{n\Pi}$. Then the set of pairs $\{c_j, O_{j\Pi} \in O_{\Pi,\mu}\}$, where $c_j \neq 0$ and their number N_1 ($N_1 \leq N$) are determined uniquely. For the remaining semicircles $O_{j\Pi} \in O_{\Pi,\mu}$ (if they exist), the values c_j are equal to zero.*

The theorem yields the following statement.

Corollary 8.5.1. *Suppose following the condition (C) holds:*

(C) *each sum $m_{j_1} + m_{j_2} + \ldots + m_{j_m} \neq 0$, where $j_1, j_2, \ldots, j_m$ $(1 \leq m \leq N)$ is an arbitrary subset from the set $1, 2, \ldots, N$.*

Then, if we set the field $u(x, y, z)$ of form (8.1.13) in the set $X_{n\Pi}$, we define uniquely the set $O_{\Pi,\mu}$, the number c_j for each $O_{j\Pi} \in O_{\Pi,\mu}$ and their number N_1 ($N_1 \leq N$).

Evidently, the condition (C) holds if, for example, all the numbers m_j have one sign. Such is the case of the Newton potential, when all the masses m_j of the points M_j are positive.

From Remark 8.5.1 follows the next one.

Remark 8.5.2. For the given values $m_j \neq 0$ and the set μ of points M_j, the number N_1 of the values c_j not equal to zero (and the number of various semicircles $O_{j\Pi}$ from the set $O_{\Pi,\mu}$) attains the maximal value N if and only if $\Pi \subseteq \Pi_\mu$. In this case $c_j = m_j$.

Remark 8.5.2 and Theorem 8.5.1 yield the following corollary.

Corollary 8.5.2. *Let $\Pi \subseteq \Pi_\mu$. Then the number N, the set $O_{\Pi,\mu}$ and the numbers m_j for each point M_j lying in $O_{j\Pi}$ for $j = 1, 2, \ldots, N$, are determined uniquely by setting the field $u(x, y, z)$ of form (8.1.13) in the set $X_{n\Pi}$.*

8.6. THE NONSTATIONARY INVERSE PROBLEM WITH THE DATA GIVEN IN THE STRAIGHT LINE FOR THE CASE 1

Let, as before, $X_{n\Pi}$ be a set consisting from $f(2n)$ various arbitrary points lying in a certain straight line Π from the plane $z = 0$. Set

$$a_j(t) = \varphi_{j_1}(t) + \varphi_{j_2}(t) + \ldots + \varphi_{j_m}(t), \tag{8.6.1}$$

where $j_1, j_2, \ldots, j_m$ are the numbers of those points $M_j \in \mu$ which lie in one semicircle $O_{j\Pi}$ from the set $O_{\Pi,\mu}$. The following theorem holds.

Theorem 8.6.1. *Suppose the number N, the coordinates of points $M_j \in \mu$ and the functions $\varphi_j(t)$, for $-\infty < t < \infty$, $j = 1, 2, \ldots, N$ are unknown. Suppose that the number n is known, so that $N \leq n$. Suppose also that the following condition holds*

(Φ) *the functions $\varphi_j(t)$ are piecewise-differentiable, and the functions $\varphi_j(t)e^{b(t)}$, where $b = \text{const} > 0$, are absolutely integrable in the axis $-\infty < t < \infty$ for $j = 1, 2, \ldots, N$ (this condition holds, for example, for finite functions $\varphi_j(t)$). Note that the property $\varphi_j(t) = 0$ for $t < t_j$ is not obligatory.*

Suppose we set the field $u(x, y, 0, t)$ of form (8.1.6) for $-\infty < t < \infty$ in the set $X_{n\Pi}$. Then the set of pairs $\{a_j(t)$ for $-\infty < t < \infty$, $O_{j\Pi} \in O_{\Pi,\mu}\}$ where $a_j(t) \not\equiv 0$ and their number N_1 ($N_1 \leq N$) is determined uniquely. For the rest semicircles $O_{j\Pi} \in O_{\Pi,\mu}$, if they exist, $a_j(t) \equiv 0$.

Proof. Introduce the Cartesian coordinate system (x, y) in the plane $z = 0$, so that the line Π coincides with a certain line $y = \text{const}$. Then the field $u(x, y, 0, t)$ of form (8.1.6) in the straight line Π is a function dependent on x:

$$u(x, y, 0, t) = u_n(x, t) = \sum_{j=1}^{N \leq n} \frac{\varphi_j(t - R_{0j}(x)/v)}{R_{0j}(x)} = \int_{-\infty}^{\infty} U_n(x, k)e^{ikt}\, dk. \tag{8.6.2}$$

In this formula, taking into account formulas (8.1.7) and (8.1.8), we have

$$U_n(x, k) = \sum_{j=1}^{N \leq n} \Phi_j(k)\, \frac{e^{-i(k/v)R_{0j}(x)}}{R_{0j}(x)}, \tag{8.6.3}$$

$$R_{0j}(x) = \sqrt{(x - x_j)^2 + r_j^2} = \{\psi_j(x)\}^{-1}, \quad r_j^2 = (y_0 - y_j)^2 + z_j^2, \tag{8.6.4}$$

where x_j, y_j, z_j are the coordinates of the point M_j; $\Phi_j(k)$ is the Fourier transform of the function $\varphi_j(t)$:

$$\varphi_j(t) = \int_{-\infty}^{+\infty} \Phi_j(k) e^{ikt}\, dk.$$

The function $u_n(x,t)$ is given for $-\infty < t < \infty$ in the set $X_{n\Pi}$; therefore, the function $U_n(x,k)$ of form (8.6.3) is determined uniquely. After combining the terms with the equal pairs (x_j, r_j) (if they exist) it is represented in the form

$$U_n(x,k) = U(x,k) = \sum_{j=1}^{\tilde{N}\le N} A_j(k) \frac{e^{-i(k/v)R_{0j}(x)}}{R_{0j}(x)}, \tag{8.6.5}$$

where

$$A_j(k) = \Phi_{j_1}(k) + \Phi_{j_2}(k) + \ldots + \Phi_{j_m}(k). \tag{8.6.6}$$

In this case, $j_1, j_2, \ldots, j_m$ $(1 \le m \le N)$ are the numbers of the points $M_j \in \mu$ with equal pairs (x_j, r_j), i. e. the points M_j lie in one semicircle $O_{j\Pi} \in O_{\Pi,\mu}$. Therefore, the function $A_j(k)$ is the Fourier transform of the function $a_j(t)$.

By the condition (Φ) of Theorem 8.6.1 (see Shilov, 1970, p. 271) the Fourier transformation $\Phi_j(k)$ of the function $\varphi_j(t)$ is not only infinitely differentiable function, but it is defined and analytical in the strip $|\operatorname{Im} k| < b$ of the complex plane k. Hence it follows that each of the functions $A_j(k)$, $U(x,k)$ is analytical in the strip $|\operatorname{Im} k| < b$. So, the following statement holds. □

Statement 8.6.1. The derivatives $A_j^{(\nu)}(k)$ of the functions $A_j(k)$ of the order ν exist for each $\nu = 0, 1, 2, \ldots$ in the axis $-\infty < k < \infty$. The functions $A_j(k)$, $j = 1, 2, \ldots, \tilde{N}$ are determined uniquely in the disk $|k| < b$ and therefore in the strip $|\operatorname{Im} k| < b$, $-\infty < k < \infty$ if we set the values $A_j^{(\nu)}(0)$, $\nu = 0, 1, 2, \ldots$. If $A_j^{\nu}(0) = 0$ for all $\nu = 0, 1, 2, \ldots$ then $A_j(k) \equiv 0$ in the disk $|k| < b$ and for $-\infty < k < \infty$.

By Statement 8.6.1, for each $\nu = 0, 1, 2, \ldots$, we define uniquely the functions $\partial^\nu U/\partial k^\nu$ for $k = 0$, $x \in X_{n\Pi}$.

It is easy to show using the induction method that for each $\nu = 0, 1, 2, \ldots$ the following formula holds:

$$\frac{\partial^\nu U}{\partial k^\nu}\Big|_{k=0} = \sum_{j=1}^{\tilde{N}\leq N} A_j^{(\nu)}(0)\psi_j(x) + \sum_{j=1}^{\tilde{N}\leq N} \{c_1 A_j^{(\nu-1)}(0)$$
$$+ c_2 A_j^{(\nu-2)}(0)\{\psi_j(x)\}^{-1} + \ldots + c_\nu A_j(0)\{\psi_j(x)\}^{-\nu+1}\}, \quad (8.6.7)$$

where $c_1, c_2, \ldots, c_{\tilde{N}}$ are known constants independent of $N, x_j, r_j, A_j^\nu(0)$. They depends only on v: $c_\nu = \alpha_\nu(-i/v)^\nu$, where the numbers α_ν are constructed by ν uniquely. The following statement holds.

Statement 8.6.2. Suppose that the functions

$$U_\nu(x) = \frac{\partial^\nu U}{\partial k^\nu}\Big|_{k=0} \quad (8.6.8)$$

are set for $\nu = 0, 1, 2, \ldots, \nu_0$ in the set $X_{n\Pi}$. Then, for each $\nu = 1, 2, \ldots, \nu_0$, the collection of triples $\{A_j^{(\nu)}(0), x_j, r_j\}$ (or, of all the pairs $\{A_j^{(\nu)}(0), O_{j\Pi} \in O_{\Pi,\mu}\}$, for which $A_j^{(\nu)}(0) \neq 0$) is determined uniquely.

Proof. First we note that the values $A_j^{(\nu)}(0)i^\nu$ and $i^\nu U_\nu(x)$ are real. This follows from the equalities

$$U(x, k) = \bar{U}(x, -k), \qquad \Phi_j(k) = \bar{\Phi}_j(-k),$$

where the bar denotes as usually the complex conjugation. Denote $A_j^{(\nu)}(0) = A_j^{(\nu)}$, $A_j(0) = A_j$. For $\nu_0 = 0$, by the function

$$U_0(x) = U(x, 0) = \sum_{j=1}^{\tilde{N}\leq N} A_j\psi_j(x)$$

(where all A_j are real), given in the set $X_{n\Pi}$, we determine (uniquely) the set of all triples $\{A_j, x_j, r_j\}$, where $A_j \neq 0$ (this can be done following Corollary 8.3.2).

For $\nu_0 = 1$, taking into account formula (8.6.7), we obtain

$$U_1(x) = \sum_{j=1}^{\tilde{N}\leq N} A_j^{(1)}\psi_j(x) + \sum_{j=1}^{\tilde{N}} c_1 A_j.$$

Since c_1 is known and all $A_j \neq 0$ are obtained, the second sum in the last inequality is known. Therefore, we can determine the sum $\sum_{j=1}^{\tilde{N}\leq N} iA_j^{(1)}\psi_j(x)$

in the set $X_{n\Pi}$, where all the values $iA_j^{(1)}$ are real. Thereby, by Corollary 8.3.2, we determine uniquely the collection of all triples $\{A_j^{(\nu)}, x_j, r_j\}$, where $A_j^{(1)} \neq 0$.

So, the statement is proved for $\nu_0 = 0, 1$. Suppose for ν_0 this statement holds, we shall prove if for $\nu_0 + 1$ (i. e., we use the induction method).

By the functions $U_\nu(x)$ given for $\nu = 0, 1, \dots, \nu_0$ in the set $X_{n\Pi}$ we define uniquely for each $\nu = 0, 1, \dots, \nu_0$ the set of triples $\{A_j^{(\nu)}, x_j, r_j\}$, i. e., the set of the pairs $\{A_j^{(\nu)}, \psi_j(x)\}$, where $A_j^{(\nu)} \neq 0$. Therefore, in the equality

$$U_{\nu_0+1}(x) = \sum_{j=1}^{\tilde{N}\leq N} A_j^{(\nu_0+1)} \psi_j(x)$$
$$+ \sum_{j=1}^{\tilde{N}} \{c_1 A_j^{(\nu_0)} + c_2 A_j^{(\nu_0-1)} \{\psi_j(x)\}^{-1} + \dots + c_{\nu_0+1} A_j \{\psi_j(x)\}^{-\nu_0}\},$$

which follows from formula (8.6.7), we define uniquely the second sum in the set $X_{n\Pi}$. This is so, because in this sum we have only those terms $\{A_j^{(\nu)}, \psi_j(x)\}$, for which $A_j^{(\nu)} \neq 0$, and the constants $c_1, c_2, \dots, c_{\nu_0+1}$ are known. Therefore, in $X_{n\Pi}$, we define the function $\sum_{j=1}^{\tilde{N}\leq N} A_j^{(\nu_0+1)} \psi_j(x)$, where all the values $i^{\nu_0+1} A_j^{(\nu_0+1)}$ are real. So, by Corollary 8.3.2, the collection of all triples $\{A_j^{(\nu_0+1)}, x_j, r_j\}$, where $A_j^{(\nu_0+1)} \neq 0$ is determined uniquely. □

So, the statement is proved for arbitrary ν_0. Hence the following statement follows.

Statement 8.6.3. Given the set of functions $\{U_\nu(x)\}_{\nu=0}^{\infty}$ of form (8.6.8) in the set $X_{n\Pi}$. In this case, the collection of all the triples $\{A_j^{(\nu)}, x_j, r_j\}$ (or the collection of all the pairs $\{A_j^{(\nu)}, O_{j\Pi} \in O_{\Pi,\mu}$, where $A_j^{(\nu)} \neq 0\}$ for each $\nu = 0, 1, 2, \dots$ and their number is determined uniquely.

From this statement the next one follows.

Statement 8.6.4. Given the set of functions $\{U(x)\}_{\nu=0}^{\infty}$ of form (8.6.8) in the set $X_{n\Pi}$; then the collection of the sets

$$\{A_j, A_j^{(1)}, A_j^{(2)}, \dots, A_j^{(\nu)}, \dots, O_{j\Pi} \in O_{\Pi,\mu}\},$$

for which for at most one value $\nu = 0, 1, 2, \dots, A_j^{(\nu)} \neq 0$, is defined uniquely and their number N_1 ($N_1 \leq N$) is defined also. For the other semicircles $O_{j\Pi} \in O_{\Pi,\mu}$, if they exist, $A_j^{(\nu)} = 0$ for all $\nu = 0, 1, \dots$.

Proof. Suppose for at least one value ν from the sequence $0, 1, 2, \ldots$ the values $A_j^{(\nu)} \neq 0$. Then, by Statement 8.6.3, for such value ν, the triple $\{A_j^{(\nu)}, x_j, r_j\}$ is determined uniquely. Therefore, the pair (x_j, r_j) exists for all values $\nu = 0, 1, \ldots$, i. e., for all sums $\sum_{j=1}^{\tilde{N}} A_j^{(\nu)} \psi_j(x)$, since the set of pairs (x_j, r_j) for all such sums (values ν) is the same. Thereby, if for some other value ν, among the pairs $\{A_j^{(\nu)}, O_{j\Pi}\}$, where $A_j^{(\nu)} \neq 0$ (they are determined uniquely following Statement 8.6.3), the semicircle $O_{j\Pi}$ (the pair x_j, r_j) is absent, then $A_j^{(\nu)} = 0$ (for this other value ν), i. e., is determined uniquely. If $O_{j\Pi}$ is among the other such pairs, then $A_j^{(\nu)} \neq 0$ and is determined uniquely also.

So, if the sequence of numbers $A_j, A_j^{(1)}, \ldots$ has at least one nonzero value $A_j^{(\nu)} \neq 0$, then, by the collection $\{U_\nu(x)\}_{\nu=0}^{\infty}$ given in $X_{n\Pi}$ we define the set $\{A_j, A_j^{(1)}, \ldots, O_{j\Pi}\}$ uniquely which was to be proved.

Now we finish the proof of Theorem 8.6.1. Suppose for a certain semicircle $O_{j\Pi}$ the function $a_j(t)$ defined by formula (8.6.1) is not equal to zero identically. Then, for at least one value $\nu = 0, 1, 2, \ldots$ the value $A_j^{(\nu)} \neq 0$ since, by Statement 8.6.1, otherwise, we should have

$$A_j(k) = \sum_{\nu=0}^{\infty} A_j^{(\nu)} \frac{k^\nu}{\nu!} \equiv 0 \quad \text{and} \quad a_j(t) = \int_{-\infty}^{\infty} A_j(k) e^{ikt} \, dk \equiv 0. \tag{8.6.9}$$

Therefore, if $a_j(t) \not\equiv 0$, by Statement 8.6.4, we define uniquely the sequence $A_j, A_j^{(1)}, \ldots$ and the semicircle $O_{j\Pi}$ (it is just the same for $a_j(t)$, $A_j(k)$, $A_j^{(\nu)}$, $\nu = 1, 2, \ldots$). By the sequence $A_j^{(\nu)}$, $\nu = 0, 1, 2, \ldots$, taking into account formulas (8.6.9), we determine the function $a_j(t)$ uniquely. For the other $O_{j\Pi} \in O_{\Pi,\mu}$, if they exist, $a_j(t) \equiv 0$. Theorem 8.6.1 is proved. □

Theorem 8.6.1 yields the following corollary.

Corollary 8.6.1. *Suppose the following condition (condition A) holds:*

(A) each sum $\varphi_{j_1}(t) + \varphi_{j_2}(t) + \ldots + \varphi_{j_m}(t) \not\equiv 0$, where $j_1, j_2, \ldots, j_m$ is an arbitrary subset from the set $1, 2, \ldots, N$.

Suppose we set the field $u(x, y, 0, t)$ of form (8.1.6) for $-\infty < y < \infty$ in the set $X_{n\Pi}$. Then, we define uniquely the set $O_{\Pi,\mu}$, the function $a_j(t)$ of form (8.6.1), $-\infty < t < \infty$, for each $O_{j\Pi} \in O_{\Pi,\mu}$ and their number N_1 ($N_1 \leq N$).

Evidently, the condition (A) holds, for example, when all $\varphi_j(t)$ have the similar sign.

Remark 8.5.1 yields the next remark.

Remark 8.6.1. Let the functions $\varphi_j(t) \not\equiv 0$ and the set μ be given. Then the number N_1 of the functions $a_j(t)$ not equal to zero identically (for $-\infty < t < \infty$) (and the number of various semicircles $O_{j\Pi}$ in the set $O_{\Pi,\mu}$) attains its maximal value (equal to N) if and only if $\Pi \subseteq \Pi_\mu$. In this case, $a_j(t) \equiv \varphi_j(t)$.

Remark 8.6.1 and Theorem 8.6.1 yield the following corollary.

Corollary 8.6.2. *Suppose $\Pi \subseteq \Pi_\mu$ is a strait line. Set the field $u(x,y,0,t)$ of form* (8.1.6), *where $\varphi_j(t) \not\equiv 0$, in the set $X_{n\Pi}$. Then, the number N, the set $O_{\Pi,\mu}$, and the function $\varphi_j(t)$ (therefore, the number t_j if $\varphi_j(t) = 0$ for $t < t_j$) for $-\infty < t < \infty$ are determined uniquely for each point M_j lying in $O_{j\Pi}$, for $j = 1, 2, \dots, N$.*

8.7. THE INVERSE STATIC AND NONSTATIONARY PROBLEMS

Inverse problem 8.1. Suppose in the static problem all the quantities defined the field $u(x,y,z)$ of form (8.1.13) are unknown: the numbers N, m_j; the coordinates x_j, y_j, z_j of the points $M_j \in \mu$ for $j = 1, 2, \dots, N$. In the set E (we shall define it below) the field $u(x,y,0)$ of form (8.1.14) is set. Respectively, suppose in the nonstationary problem, the following quantities defined the field $u(x,y,z,t)$ of form (8.1.4) or (8.1.5) are unknown: the number N, the coordinates x_j, y_j, z_j of the points $M_j \in \mu$ and the functions $\varphi_j(t)$ for $-\infty < t < \infty$ (so, the numbers t_j such that $\varphi_j(t) = 0$ for $t < t_j$ are unknown also). In the set E the field $u(x,y,0,t)$, $-\infty < t < \infty$ of form (8.1.6) is given and the number v is given also.

Let Π be an arbitrary straight line in the plane $z = 0$. We shall denote by $X_{n\Pi}, S_\Pi, P_\Pi, R_\Pi$ the set lying in the line Π so that

$X_{n\Pi}$ consists of $f(2n)$ arbitrary points, $f(n) = n\{n(n^2-1)/2 + 1\}$;

P_Π is an arbitrary sequence with a finite limit point;

S_Π is an arbitrary sequence with the infinite limit point;

R_Π is an arbitrary segment of extremely small length.

The set E consists only of the points which belong to the plane $z = 0$ and is one of the following sets (both in static and in nonstatic problems):

E_{01} is the union of two sets of points $X_{n\Pi_i}$, where $i = 1, 2$. The straight lines Π_1, Π_2 are arbitrary but not coincide;

E_{02} is the sum of $(C_n^2 + 2)$ sets of points $X_{n\Pi_i}$, where the straight lines Π_i form the set π_2 of $(C_n^2 + 2)$ arbitrary not coinciding straight lines;

E_{03} is the union of $(C_{2n}^2 + 1)$ sets of points $X_{n\Pi_i}$. The straight lines Π_i form the set π_3 consisting of $(C_{2n}^2 + 1)$ arbitrary various straight lines;

E_s, $s = 1, 2, 3$ is the set obtained from the set E_{0s} by the change of $X_{n\Pi_i}$ by P_{Π_i};

E_s', $s = 1, 2, 3$ is the set obtained from the set E_{0s} by the change of $X_{n\Pi_i}$ by S_{Π_i} or R_{Π_i};

E_4 is the sum of sets P_{Π_i}, where the straight lines Π_i form the countable sequence of various straight lines;

E_4' is the set obtained from E_4 by the change of P_{Π_i} by S_{Π_i} or by R_{Π_i};

E_5 is the circle of extremely small radius and with the center chosen arbitrarily;

E_6 is an arbitrary bounded domain.

Each of the sets E_s, where $s = 1, 2, 3, 4$ is countable. The number n is such that $N \leq n$ is given only in the cases $E = E_{0s}$, E_s, E_s', $s = 1, 2, 3$. In the cases $E = E_{01}$, E_1, E_1' we set that the condition (C) holds in the static problem and the condition (A) holds in the nonstationary problem. (These conditions are formulated in Corollary 8.5.1 and Corollary 8.6.1. Besides, in the nonstationary problem we assume always that the condition (Φ) is imposed on the functions $\varphi_j(t)$ (see the formulations of Theorem 8.6.1).

In the static problem, we need to find the numbers N, m_j and the coordinates of the points M_j for each $j = 1, 2, \ldots, N$. In the nonstationary problem we need to find the number N, the coordinates of the points M_j and the functions $\varphi_j(t)$ for $-\infty < t < \infty$ (therefore, the numbers t_j if the functions satisfy the property: $\varphi_j(t) = 0$ for $t < t_j$) for each $j = 1, 2, \ldots, N$.

Note that the functions $\varphi_j(t)$, generally speaking, may have not the property: $\varphi_j(t) = 0$ for $t < t_j$ (similar as in Theorem 8.6.1).

The following theorem holds.

Theorem 8.7.1. *In each of the cases* $E = E_{03}$, E_3, E_3', E_4, E_4', E_5, E_6 *inverse problem 8.1 has a unique solution* $\{N, M_j, m_j (j = 1, 2, \ldots, N)\}$ *in the static case and a unique solution* $\{N, M_j, \varphi_j(t)$ *for* $-\infty < t < \infty$, $t_j (j = 1, 2, \ldots, N)\}$ *in the nonstationary case.*

In each of the cases $E = E_{01}$, E_{02}, E_1, E_2, E_1', E_2' *the following results hold.*

First, we can define whether the pair Π_1, Π_2 *has the property* B_μ *or not. (In the cases* $E = E_{02}$, E_2, E_2', *we denote by* Π_1, Π_2 *two arbitrary lines from the set* π_2, *which not belong to the set* Π_μ*). If the property* B_μ *holds, then inverse problem 8.1 has a unique solution.*

If the property B_μ *does not hold, we define uniquely the sets* $O_{\Pi_1,\mu}$, $O_{\Pi_2,\mu}$, *the finite set of point* $\mu_{1,2}$ *which contains the set* μ, *the values* c_j *(in the static problem), the functions* $a_j(t)$ *for* $-\infty < t < \infty$ *(in the nonstationary problem) for all semicircles* $O_{j\Pi_1}$ *and* $O_{j\Pi_2}$. *Moreover, in the cases* $E = E_{02}$, E_2, E_2', $a_j(t) \equiv \varphi_j(t)$ *and the numbers* N *and* t_j *are determined uniquely. Each point of the set* $\mu_{1,2}$ *lying in a semicircle from* $O_{\Pi_1,\mu}$ *or* $O_{\Pi_2,\mu}$ *which does not contain another points from* $\mu_{1,2}$ *is a point from the set* μ. *For this point, respectively, the number* m_j *and the function* $\varphi_j(t)$ *for* $-\infty < t < \infty$ *(and the number* t_j*) are determined uniquely.*

Besides, by the set $\mu_{1,2}$, *we can always construct the following set* $\Pi_{\mu_{1,2}}$ *consisted of* $C_{\bar N}^2$ *straight lines at the plane* $z = 0$ *(*$\bar N$ *is a number of points in* $\mu_{1,2}$, $\bar N \le N^2$*). The additional given field* $u(x, y, 0)$ *of form* (8.1.14) *or the field* $u(x, y, 0, t)$ *for* $-\infty < t < \infty$ *of form* (8.1.6) *in the set* $X_{n\Pi_3}$ *in the cases* $E = E_{01}$, E_{02}, *in the set* P_{Π_3} *in the cases* $E = E_1, E_2$, *in the set* S_{Π_3} *or* R_{Π_3} *in the cases* $E = E_1', E_2'$, *where the straight line* $\Pi_3 \overline{\in} \Pi_{\mu_{1,2}}$ *is arbitrary, allows us to find uniquely all the desired values.*

Proof. First we consider the case of static problem.

The case $E = E_{01}$. From the Corollary 8.5.1 it follows that given the set $X_{n\Pi_i}$, we determine uniquely the set $O_{\Pi_i\mu}$ and the values c_j for all semicircle $O_{j\Pi_i} \in O_{\Pi_i,\mu}$ $(i = 1, 2)$ and, therefore, the set $\mu_{1,2}$. By the set $\mu_{1,2}$, we can say whether the lines Π_1 and Π_2 satisfy the property B_μ.

Suppose Π_1 and Π_2 satisfy the property B_μ. Then $\mu_{1,2} = \mu$ (see § 8.2) and the coordinates of all points $M_j \in \mu$ and their number N are found. Now we derive the values m_j. Take the point M_j. By the property B_μ, for at least one of two semicircles $O_{j\Pi_1}$ and $O_{j\Pi_2}$ which pass through the point M_j (suppose, for definiteness, it is $O_{j\Pi_1}$) there are no points from the set μ; therefore, the value c_j corresponding to $O_{j\Pi_1}$ is equal to m_j.

Assume now that the property B_μ does not hold. Then the set $\mu_{1,2}$ may have the points which do not belong to μ. In this case we define the set $\mu_{1,2}$

which contains the desired set μ. As a result we obtain the estimate from above and from below for the number N:

$$\max_i N_{1i} \le N \le \bar{N},$$

where N_{1i} is the number of semicircles in the set $O_{\Pi_i,\mu}$; $\bar{N}$ is the number of points in $\mu_{1,2}$, $\bar{N} \le N^2$ [ii]. We find also the values c_j for each semicircle $O_{j\Pi_i} \in O_{\Pi_i,\mu}$ $(i = 1, 2)$. In this case, if in a certain semicircle $O_{j\Pi_i}$, only one point from the set $\mu_{1,2}$ lies, then each such point belongs to the set μ, and for this point, the number m_j is determined uniquely (since $m_j = c_j$, where c_j corresponds to $O_{j\Pi_i}$).

For the case $E = E_{01}$ we shall prove the last part of Theorem 8.7.1. To this end, construct on the plane $z = 0$ the set of lines $\Pi_{\mu_{1,2}}$ (see § 8.2, in this case $\mu_{1,2}$ plays the role of μ). The set $\Pi_{\mu_{1,2}}$ is finite and consists of no more than $C^2_{\bar{N}}$ various lines. We take in the plane $z = 0$ any line $\Pi_3 \,\overline{\in}\, \Pi_{\mu_{1,2}}$. Then, each semicircle with the center in Π_3, lying in the plane normal to Π_3 contains no more than one point from the set $\mu_{1,2}$. Set in Π_3 the set of points $X_{n\Pi_3}$ and the field $u(x, y, 0)$ of form (8.1.14) in the set $X_{n\Pi_3}$. As a result we define uniquely in $X_{n\Pi_3}$ (by the Corollary 8.5.2) the set $O_{\Pi_3,\mu}$ and the correspondent numbers $c_j = m_j$. Only those points from the set $\mu_{1,2}$ through which come the semicircles from the set $O_{\Pi_3,\mu}$ select the set μ: $\mu = \mu_{1,2} \cap O_{\Pi_3,\mu}$.

The case $E = E_{02}$. In this case the set π_2 contains at least two straight lines which are not contained in the set Π_μ (see § 8.2). For each straight line $\Pi_i \in \pi_2$, by Theorem 8.5.1, we define uniquely the number N_1 of the values c_j not equal to zero. By the Remark 8.5.2, those lines for which N_1 attains the maximal value (their number is not less than two) do not belong to Π_μ. Choose from the set π_2 two arbitrary such lines Π_1 and Π_2. For each of them, by Corollary 8.5.2, we define uniquely the set $O_{\Pi_i,\mu}$, the numbers $N_1 = N$ and m_j and find the set $\mu_{1,2} = O_{\Pi_1,\mu} \cap O_{\Pi_2,\mu}$. Further we prove similar as for the case $E = E_{01}$: define the set $\Pi_{\mu_{1,2}}$, set the field $u(x, y, 0)$ in $X_{n\Pi_3}$, where $\Pi_3 \,\overline{\in}\, \Pi_{\mu_{1,2}}$ and so on.

The case $E = E_{03}$. Suppose that there are two sets

$$\{N, \{m_j\}_1^N, \mu = \{M_j\}_j^N\} \text{ and } \{N', \{m_j\}_1^{N'}, \mu' = \{M'_j\}_j^{N'}\}, \tag{8.7.1}$$

[ii] The maximum of the number of points in $\mu_{1,2}$ (equal to N^2) is achieved when each semicircle $O_{j\Pi_1} \in O_{\Pi_1,\mu}$ is intersected with each semicircle $O_{j\Pi_2} \in O_{\Pi_2,\mu}$.

Example: All the points $M_j \in \mu$ lie in the semisphere, and the lines Π_1 and Π_2 are passing through its center.

where $N \le n$, $N' \le n$ generated the fields $u(x,y,z)$ and $u'(x,y,z)$ of form (8.1.13) respectively. The union of the sets μ and μ' we denote by μ^*: $\mu^* = \mu \cup \mu'$. Then the number of points in μ^* is not greater than $2n$ and the difference $u_{2n} = u - u'$ is the field of form (8.1.13) also:

$$u_{2n}(x,y,z) = \sum_{j=1}^{N} m_j \Big\{ \frac{1}{R_j(x,y,z)} + \frac{\rho}{R_j(x,y,-z)} \Big\} + \sum_{j=1}^{N'} (-m'_j) \Big\{ \frac{1}{R'_j(x,y,z)} + \frac{\rho}{R'_j(x,y,-z)} \Big\}, \quad (8.7.2)$$

In the plane $z = 0$ this difference is the field of form (8.1.14):

$$u_{2n}(x,y,0) = \sum_{j=1}^{N} \frac{m_j}{R_{0j}} + \sum_{j=1}^{N'} \frac{-m'_j}{R'_{0j}},$$

where the values R'_j, R'_{0j} are obtained from R_j, R_{0j} by the change of x_j, y_j, z_j by x'_j, y'_j, z'_j (the coordinates of the points M'_j).

The function u_{2n} is equal to zero in the set $X_{n\Pi_i}$, where $i = 1, 2, \ldots, C^2_{2n} + 1$ (i.e., in the set E_{03}). Since the number of lines in the set Π_{μ^*} (see, § 2, in this case μ plays the role μ^*) is not greater than C^2_{2n}, then, in the set π_3, there exists at least one straight line $\Pi \overline{\in} \Pi_{\mu^*}$. Therefore, by Corollary 8.3.1 or by Theorem 8.8.2, formulated below (in this case, n plays the role $2n$, and the role μ plays μ^*), we obtain that $u_{2n}(x,y,z) \equiv 0$. Hence, we obtain

$$u(x,y,z) \equiv u'(x,y,z);$$

therefore, $u(x,y,0) = u'(x,y,0)$ in the whole plane $z = 0$. However, the plane $z = 0$ contains always a set of the type E_{02} and the additional set $X_{n\Pi_3}$ from the case $E = E_{02}$. Therefore, both our sets (8.7.1) coincide (otherwise, we obtain the contradiction). This means that all the desired values in the inverse problem 8.1 are determined uniquely for $E = E_{03}$.

We shall demonstrate also another proof (after obtaining the identity $u_{2n} \equiv 0$), which does not use referring to the case $E = E_{02}$. Assume that both sets (8.7.1) does not coincide. Then, for at least one pair m_j, M_j or m'_j, M'_j the equal pair does not exist. This means that the sum for the functions u_{2n} in expression (8.7.2) contains the term $\tilde{m}_j / \tilde{R}_j$, where $\tilde{m}_j \neq 0$ and $\tilde{R}_j$ is one of the values R_j, R'_j. Suppose $\tilde{R}_j$ corresponds to the point $\tilde{M}_j$ ($\tilde{M}_j$ is one point in μ or μ'). Then, $\vec{R}_j \to 0$ for $(x,y,z) \to \tilde{M}_j$; therefore, the term $\tilde{m}_j/\tilde{R}_j$ tends to infinity, while all other terms in the sum for u_{2n} of form (8.7.2) are bounded in a neighborhood of the point $\tilde{M}_j$. This means that $u_{2n} \to \infty$ for $(x, y, z) \to \tilde{M}_j$ which contradicts to the identity $u_{2n} \equiv 0$. Therefore, sets (8.7.1) coincide, and the inverse problem 8.1 has a unique solution in the case $E = E_{03}$.

Another cases. Each countable infinite set of points, or each segment in each line Π from the plane $z = 0$ contains the set $X_{n\Pi}$ for any natural n and $N \leq n$. So, the sets E_s, E'_s, $s = 1, 2, 3$ contain the set E_{0s}. Therefore, the case $E = E_s$, E'_s $(s = 1, 2, 3)$ is reduced to the case $E = E_{0s}$ researched above. In this case, we define uniquely the same values that in the case $E = E_{0s}$ (if we assume the contrary, we obtain the contradiction).

In the cases $E = E_4$, E'_4, E_5, E_6, the set E contains the set of the type E_{03} for each natural N and $n \geq N$. Hence follows that Theorem 8.7.1 holds for the static problem. □

The proof of Theorem 8.7.1 for the nonstationary problem is analogous. It is obtained from the proof above by the change of the words "the field $u(x, y, 0)$ of form (8.1.14)" and "the number c_j", "the number m_j" by the words "the field $u(x, y, 0, t)$ of form (8.1.6)", "the function $a_j(t)$ for $-\infty < t < \infty$", "the function $\varphi_j(t)$ for $-\infty < t < \infty$ and the numbers t_j". The words "Corollary 8.5.1", "Theorem 8.5.1", "Remark8.5.2", "Corollary 8.5.2" should be replace by the words "Corollary 8.6.1", "Theorem 8.6.1", "Remark8.6.1", "Corollary 8.6.2".

8.8. ON ZEROS OF THE FIELD $u(x, y, z)$ OF FORM (8.1.13)

Denote by $X'_{n\Pi}$ the set consisted of no less than $f(n)$ different points in an arbitrary straight line Π lying in the plane $z = 0$.

Corollaries 8.2.1, 8.3.1 yield the following theorem.

Theorem 8.8.1. *Let $n \geq N$ be natural and Π be an arbitrary straight line in the plane $z = 0$. If the field $u(x, y, 0)$, of form (8.1.14) is equal to zero in the set $X_{n\Pi}$, then for all semicircles $O_{j\Pi} \in O_{\Pi,\mu}$, we have $c_j = 0$; therefore, $u(x, y, 0) = 0$ in the line Π. If, in this case, $\Pi \bar{\in} \Pi_\mu$, then $c_j = m_j = 0$ for $j = 1, 2, \dots, N$ and $u(x, y, z) \equiv 0$.*

From this theorem the following statement follow.

Corollary 8.8.1. *Suppose the set of numbers $\{m_j\}_1^N$ satisfies the condition (C) (see § 4, Corollary 8.5.1). Then there does not exist such collection $\{\Pi, N, \mu, \{m_j\}_1^N\}$ for which $u(x, y, 0) = 0$ in $X'_{n\Pi}$ and the more so as in $X_{n\Pi}$ or in the set E defined in the inverse problem 8.1. This means that the* *function $u(x,y,0)$ has no more than $\{f(n)-1\}$ zeros in each straight line Π of the plane $z=0$.*

Theorem 8.8.2. *Let E_n^* be the union of the sets $X'_{n\Pi_i}$ ($i = 1,2,\dots,C_n^2+1$), where the straight lines Π_i lie in the plane $z=0$ and $\Pi_i \neq \Pi_j$, $i \neq j$. Then if $u(x,y,0) = 0$ in the set E_n^*, then $m_j = 0$ for all $j = 1,2,\dots,N$ and $u(x,y,z) \equiv 0$.*

Corollary 8.8.2. *Suppose Π_1 and Π_2 be two not coinciding straight lines in the plane $z=0$ and the set μ belongs to the class of sets consisted from $N \leq n$ points, where the pair Π_1, Π_2 satisfies the property B_μ. Then, if $u(x,y,0) = 0$ in the sets $X'_{n\Pi_1}$ and $X'_{n\Pi_2}$, then $m_j = 0$ for $j = 1,2,\dots,N$ and $u(x,y,z) \equiv 0$.*

8.9. THE ZEROS OF THE FUNCTION $u(x,y,z,t)$ OF FORM (8.1.4) OR (8.1.5) IN THE PLANE $z=0$

Definition 8.9.1. We shall call the point (x_0,y_0) by t-zero of the function $u(x,y,0,t)$ if $u(x_0,y_0,0,t) = 0$ for $-\infty < t < \infty$.

Further, we suppose that the condition (Φ) holds formulated in Theorem 8.6.1.

Corollaries 8.2.1, 8.3.1, Statement 8.6.1 and formula (8.6.7) yield the following theorem.

Theorem 8.9.1. *Let $n \geq N$ be an arbitrary natural number and Π be a straight line in the plane $z=0$. If the function $u(x,y,0,t)$ of form (8.1.6) has no less than $f(n)$ t-zeros in the line Π (i. e., if the field $u(x,y,0,t)$ of form (8.1.6) is equal to zero for $-\infty < t < \infty$ in the set $X'_{n\Pi}$), then $u(x,y,0,t) \equiv 0$ in Π and the function $a_j(t) \equiv 0$ for each semicircle $O_{j\Pi} \in O_{\Pi,\mu}$. If, in this case, $\Pi \overline{\in} \Pi_\mu$, then $u(x,y,z,t) \equiv 0$ and $a_j(t) = \varphi_j(t) \equiv 0$ for $j = 1,2,\dots,N$.*

In other words, the function $u(x,y,0,t)$ either has no more than $\{f(n)-1\}$ t-zeros in each straight line Π of the plane $z=0$ or $u(x,y,0,t) \equiv 0$.

The following statements follow from Theorem 8.8.2.

Corollary 8.9.1. *Suppose the set of functions $\{\varphi_j(t)\}_1^N$, satisfies the condition (Φ). Then there exists no such collection $\{\Pi, N, \mu, \{\varphi_j(t)\}_1^N\}$ so that the field $u(x,y,0,t)$ of form (8.1.6) has t-zeros in the set $X'_{n\Pi}$. This means that under the condition Φ the function $u(x,y,0,t)$ has no more than $\{f(n)-1\}$ t-zeros in each straight line Π in the plane $z=0$.*

Theorem 8.9.2. *Suppose the field $u(x,y,0,t)$ of form (8.1.6) has no less than $f(n)$ t-zeros in each of (C_n^2+1) arbitrary various straight lines Π_i from the plane $z=0$, $i=1,2,\dots,C_n^2+1$. This means that the function $u(x,y,0,t)$ has t-zeros in the set E_n^* defined in Theorem 8.8.2. Then $\varphi_j(t)\equiv 0$ for $j=1,2,\dots,N$ and $u(x,y,z,t)\equiv 0$.*

Corollary 8.9.2. *Suppose the conditions of Corollary 8.9.1 hold. Then, if the function $u(x,y,0,t)$ of form (8.1.6) has t-zeros in the sets $X'_{n\Pi_1}$ and $X'_{n\Pi_2}$, then $\varphi_j(t)\equiv 0$ for $j=1,2,\dots,N$ and $u(x,y,z,t)\equiv 0$.*

8.10. SOLUTION OF THE NONSTATIONARY INVERSE PROBLEM 8.1 IN THE CASE 2 FOR $E=E_1,E_2,E_3,E_4,E_5,E_6$

In the cases $E=E_1,E_2,E_3,E_4$ the field $u(x,y,0,t)$ of form (8.1.6) in each straight line Π at the plane $z=0$ which belongs to the set of lines π_2, π_3[iii)] etc. is set on the sequence P_Π of arbitrary points from this line converging to a certain finite limit point. We can suggest the way of determination of the values x_j, r_j, $a_j(t)$ or $\varphi_j(t)$ corresponding to this line Π, by giving the field $u(x,y,0,t)$ of form (8.1.6) set in the sequence P_Π. It is based on the analytical extension of the function $U(x,y,k)$ of form (8.1.8) (the Fourier transform of the function $u(x,y,0,t)$ with respect to the spatial variable x). We determine the values x_j, r_j, $a_j(t)$ or $\varphi_j(t)$ for each of the straight lines Π_i entering the set of lines π_2, π_3 etc. for the cases $E=E_s$ $(s=1,2,3,4)$. Then we solve the inverse problem 8.1 following the scheme of § 8.7. After, we construct the set $\mu_{1,2}$ and so on. The set E'_s consisted of the segments R_{Π_i} of extremely small length contains the set of the type E_s $(s=1,2,3,4)$, and the sets E_5, E_6 contain the set of the type E_3. Therefore, in this section, we shall consider only the problem of determination of the values $x_j, r_j, a_j(t)$ or $\varphi_j(t)$ given the field $u(x,y,0,t)$ of form (8.1.6) for one sequence $P_\Pi\in\Pi$.

8.10.1. Extension of the function $U(x,y_0,k)$ into the complex plane by the spatial variable. For a certain $j=1,2,\dots,N$, we consider a function

$$w_j(z)=f\{(z-x_j)^2+r_j^2\},$$

where $r_j=\{(y_0-y_j)^2+h_j^2\}^{1/2}>0$ ($y_0\in(-\infty,\infty)$ is fixed), $h_j>0$, $h_j=z_j$, $z=x+ir$ is a complex variable (i is the imaginary unit), $f(\zeta)$ is those branch

[iii)] See the formulation of the inverse problem 8.1

of two-valued function $\sqrt{\zeta}$ for which $f(\zeta) > 0$ for $\zeta > 0$ (Figure 8.2a). Denote by γ_j the segment $-r_j \le \operatorname{Im} z \le r_j$ of the line $\operatorname{Re} z = x_j$, and denote by L_1 and L_2 the rays $\operatorname{Im} z > r_j$ and $\operatorname{Im} z < -r_j$ of this line. It is easy to show that the function $w_j(z)$ satisfies the conditions of symmetry principle of Riemann—Schwartz for conformal mappings in the half-plane $\operatorname{Re} z \ge x_j$ (see Lavrentiev and Shabat, 1973, p. 158). Therefore, this function can be extended analytically through the rays L_1 and L_2 in the half-plane $\operatorname{Re} z < x_j$. We denote this extension by $w_j(z)$.

The function $w_j(z)$ (together with its analytical extension) is analytical and univalent in the exterior of the segment γ_j, which we denote by D_j. This function maps conformally (according to the symmetry principle) the domain D into the exterior of the segment $[-r_j, r_j]$ of the axis $\operatorname{Im} w_j = 0$ (Figure 8.2b). In all the points of the segment γ_j the function $w_j(z)$ is not analytical. Namely, if we pass through each point of the cut γ_j (except its ends), the function $w_j(z)$ has the gap. However, in each of half-planes $\operatorname{Re} z \ge x_j$, and $\operatorname{Re} z \le x_j$ the function $w_j(z)$ is continuous. The zeros of $w_j(z)$ lie only in the points $z = x_j \pm i r_j$. Besides, by the symmetry principle, for each $y_0 = \text{const} \in (-\infty, \infty)$ and $j = 1, 2, \dots, N$, we have

$$R_{0j}(x, y_0) = \begin{cases} w_j(x), & \text{if} \quad x \ge x_j, \\ -w_j(x), & \text{if} \quad x \le x_j. \end{cases}$$

Therefore for each fixed values $k \in (-\infty, \infty)$ and $y_0 \in (-\infty, \infty)$, each of $(N+1)$ functions dependent on the complex variable $z = x + ir$ of the form

$$U(z, y_0, k) = U_{j_0}(z, y_0, k) = \sum_{j \le j_0} \Phi_j(k) \frac{e^{-i(k/v) w_j(z)}}{w_j(z)} - \sum_{j > j_0} \Phi_j(k) \frac{e^{i(k/v) w_j(z)}}{w_j(z)} \tag{8.10.1}$$

has the following properties.

1) For real $z = x$ in the segment $x_{j_0} \le x \le x_{j_0} + 1$, $j_0 = 0, 1, \dots, N$ we have

$$U(z, y_0, k) = U_{j_0}(z, y_0, k) = U(x, y_0, k),$$

where the function $U(x, y_0, k)$ is defined by formula (8.1.8). We set here $x_0 = -\infty$; $x_{N+1} = \infty$ and the coordinates x_j of the pairs M_j $(j = 1, 2, \dots, N)$ are numerated by nondecreasing, i. e. $x_1 \le x_2 \le \dots \le x_N$.

2) The function $U(z, y_0, k)$ is analytical and univalent in $(n+1)$-connected domain $D = \bigcap_{j=1}^{N} D_j$ which is an open complex plane z with the set of parallel cuts by the segments γ_j $(j = 1, 2, \dots, n \le N)$ and is continuous in D up to each side of each cut γ_j except for, may be, their

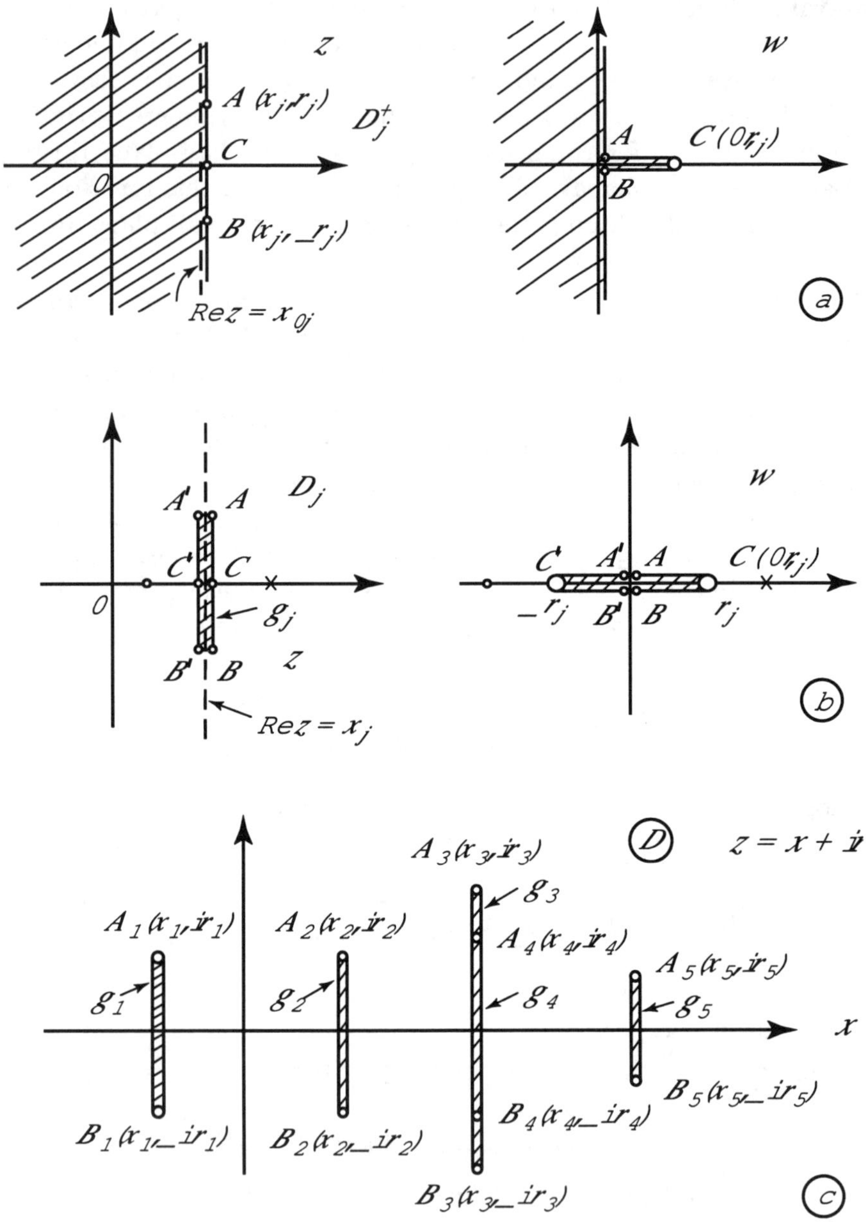

Figure 8.2. To the analytical extension of the function $U(x, y, k)$ to the complex plane with respect to the spatial variable:
(a) the mapping $w = w_j(z)$ for $\operatorname{Re} z > x_j$; (b) the mapping $w = w_j(z)$ of the domain D_j; (c) the domain D where $U(z, y, k)$ is analytical. In this case $N = 5$, $N_\gamma = 4$. The segment γ_3 contains the point $z = x_4 \pm i r_4$ inside itself

ends $z_j = x_j \pm ir_j$ (Figure 8.2c). The number n of these cuts may be less than N if in the right-hand side of equality (8.10.1) are the terms with similar values x_j. The segments γ_j for such values x_j are laid on each other. This situation corresponds to the case when in one plane $x = x_j$ there lies several points M_j.

3) The function $U(z, y_0, k)$ may have infinite limit values only in the points $x_j \pm ir_j$ but it is not necessarily. Suppose that, in the sum (8.10.1), after combining the terms with similar pairs of numbers (x_j, r_j), the coefficient before the function $e^{-i(k/v)w_j(z)}/w_j(z)$ (equal to the sum of form (8.6.6))

$$A_j(k) = \Phi_{j_1}(k) + \Phi_{j_2}(k) + \ldots + \Phi_{j_m}(k), \tag{8.10.2}$$

where $j_1, j_2, \ldots, j_m$ $(1 \leq m \leq N)$ are the numbers of the points M_j with similar pairs (x_j, r_j) is not equal to zero. Then, in the point $x_j \pm ir_j$, the function $U(z, y_0, k)$ has the infinite limit value (when tending to this point along any curve lying in D). If this coefficient is equal to zero, then, the function $U(z, y_0, k)$ is bounded in the point $x_j \pm ir_j$, and the cut γ_j is absent. Each point, where $U(z, y_0, k)$ has the infinite limit value, may correspond to several points M_j (if in (8.10.1) there are the terms with equal pairs (x_j, r_j)). Such situation holds when certain points M_j lie not only in one plane $x = x_j$, but in one circle of the radius r_j with the center in the line $y = \text{const}$.

Thus, the function $U(x, y_0, k)$ admits the analytical extension with respect to the variable x from each of the segments $[x_j, x_{j+1}]$ of the real axis into the complex domain D. (The analogous extension exists evidently with respect to the variable y for fixed x and k). In this case, the coordinates of the points M_j are connected with disposition of complex singular points of the function $U(z, y_0, k)$. This connection helps us to investigate the inverse problem.

Without loss of generality, we may assume that in the expression (8.1.8) for the function $U(x, y_0, k)$ and in the expression (8.10.1) for the function $U(z, y_0, k)$ we combine the terms with equal pairs (x_j, r_j), if they are, analogous to (8.6.5), (8.6.6), i. e.,

$$U(z, y_0, k) = U_{j_0}(z, y_0, k) = \sum_{j \leq j_0} A_j(k) \frac{e^{-i(k/v)w_j(z)}}{w_j(z)} - \sum_{j > j_0} A_j(k) \frac{e^{i(k/v)w_j(z)}}{w_j(z)}, \tag{8.10.3}$$

where $A_j(k)$ are defined by formula (8.10.2).

Evidently, we shall have the cuts γ_j and the singular points $x \pm ir_j$ of the function $U(z, y_0, k)$ for those and only those values j, for which $A_j(k) \not\equiv 0$.

8.10.2. Determination of the values $x_j, r_j, A_j(t)$, or $\varphi_j(t)$ by giving the data in P_Π. Suppose $P_\Pi = \{p_n\}$ is an infinite sequence of points of the plane $z = 0$ lying in one straight line Π, converging to a certain finite point $\bar{p}$. We shall say that the sequence P_Π has (relatively the set μ) the property A_μ if in the line, normal to Π, passing through the limit point $\bar{p}$ and lying in the plane $z = 0$, there are no projections of the points μ onto the plane $z = 0$.

Suppose that the numbers N, v, the coordinates x_j, y_j, z_j of the points M_j and the functions $\varphi_j(t)$ for $-\infty < t < \infty$, $j = 1, 2, \ldots, N$, are unknown (so the numbers t_j are unknown). We know the field $u(x, y, 0, t)$ of form (8.1.6) in the sequence of points P_Π.

Choose such a coordinate system x, y in the plane $z = 0$, that the line Π coincide with a line $y = y_0 = \text{const}$.

By formula (8.1.9) we define uniquely the function $U(x, y_0, k)$ for $-\infty < k < \infty$ in the sequence of points P_Π. Note that there exists such fixed value k, that for each $j = 1, 2, \ldots, N_1$, $(N_1 \le N)$, the value $A_j(k) \ne 0$. Up to formula (8.10.4) we use this meaning of k.

Thus, the complex function $U(z, y_0, k)$ determined by formulas (8.10.1) or (8.10.3) is defined uniquely in a certain sequence Z of the points $z_n = x_n$ $(n = 1, 2, \ldots)$ of the real axis of the complex plane $z = x + ir$. Since the sequence P_Π satisfies the property A_μ, the limit point $\bar{x} = \lim_{n\to\infty} x_{1n}$ of the sequence Z does not coincide with any of the coordinates x_j of the points M_j $(j = 1, 2, \ldots, N)$ in the chosen coordinate system x, y, z. Therefore, the limit point $\bar{x}$ is the interior point of the domain D where the function $U(z, y_0, k)$ is analytical (see, section 10.1). By the uniqueness theorem for analytical functions (Lavrentiev and Shabat, 1973, p. 73) the function $U(z, y_0, k)$ is defined in the domain D uniquely. Therefore, the coefficients c_n of Taylor's expansion of the function $U(z, y_0, k)$ in the point $\bar{x}$ are determined uniquely. Also, we can derive by the Cauchy—Hadamard formula

$$\frac{1}{\rho} = \overline{\lim_{n\to\infty}} \sqrt[n]{|c_n|}$$

the convergence radius ρ where $\overline{\lim}$ denotes the upper limit. It is known that ρ coincides with the distance from the point $\bar{x}$ to the point nearest to $\bar{x}$ where the analytic property of $U(z, y_0, k)$ fails. So this is the distance to the nearest cut γ_i. Therefore, such cut lies in one of the lines $\operatorname{Re} z = \bar{x} \pm \rho$, or in both lines. In order to select the plane where the cut lies, we construct the Taylor series for the function $U(z, y_0, k)$ with the center in two points a_1

and a_2 of real axis, so that

$$\bar{x} - \rho < a_1 < \bar{x} < a_2 < \bar{x} + \rho,$$

and derive their radii of convergence ρ_1 and ρ_2. If ρ_m coincides with the distance from the point a_m to $x_0 = \bar{x} + (-1)^m \rho$ $(m = 1, 2)$, then the cut γ_j nearest to $\bar{x}$ is disposed in the line $\operatorname{Re} z = x_0$.

Then we derive the limit values

$$\lim_{z \to x_0 + ir} U(z, y, k), \quad (z \in D),$$

and, therefore, the points $x_0 + ir$ where these values are infinite. The points with maximal value $|r|$ correspond to the ends of the cut in question. Extending the function $U(z, y_0, k)$ analytically using power series along the chains of circles to the left and to the right from the point $\bar{x}$ and enveloping the cut that we found, we can derive all other cuts γ_j $(j = 1, 2, \ldots, N_\gamma \leq N_1)$ and all the points $z = x_j \pm ir_j$ $(j = 1, 2, \ldots, N_1,\ N_\gamma \leq N_1 \leq N)$, where the function $U(z, y_0, k)$ has infinite limit values.

Therefore, we determine uniquely all the numbers

$$x_j, r_j = \{(y_0 - y_j)^2 + h_j^2\}^{1/2},$$

where x_j, y_j, h_j are the coordinates of the desired points $M_j \in \mu$, for which $A_j(k) \neq 0$.

Take a pair of numbers x_j, r_j. Evidently, in the half-space $z \geq 0$, the geometric locus of points with coordinates satisfying the equalities

$$x = x_j, \qquad \{(y_0 - y)^2 + z^2\}^{1/2} = r_j$$

is the semicircle $O_{j\Pi}$ lying in the plane $x = x_j$ normal to the straight line Π, with the center in Π and the radius r_j. Thus, given the field $u(x, y, 0, t)$ in the sequence of points P_Π, we find all semicircles $O_{j\Pi} \in O_{\Pi,\mu}$, where $A_j(k) \neq 0$.

We find now the functions $a_j(t)$ of form (8.6.1). Consider the complex plane $z = x + ir$. Since the singular point $z_j = x_j + ir_j$ correspondent to the point $M_j \in \mu$ is independent of k, we define the function $A_j(k)$ for $-\infty < k < \infty$ by the formula

$$A_j(k) = \pm \lim_{z \to z_j} \{w_j(z) U(z, y, k)\}, \quad z \in D. \tag{8.10.4}$$

Here the sign $(+)$ holds for $x_j < \bar{x}$ and the sign $(-)$ holds when $x_j > \bar{x}$. Then we find the function $a_j(t)$ for $-\infty < t < \infty$:

$$a_j(t) = \int_{-\infty}^{\infty} A_j(k)e^{ikt}\,\mathrm{d}k, \quad j = 1, 2, \ldots, N_1 \leq N.$$

If $\Pi \in \Pi_\mu$, then $A_j(k) = \Phi_j(k)$, $a_j(t) = \varphi_j(t)$ and we solve further the inverse problem 8.1 following the scheme of § 8.7.

8.11. STATIONARY INVERSE PROBLEM

Inverse problem 8.2. Suppose that all the values: the numbers N, v, k_j, A_j, θ_j and the coordinates of the points M_j, $j = 1, 2, \ldots, N$ which define the field $u(x, y, z, t)$ of form (8.1.10) or (8.1.11) are unknown. The field u of form (8.1.10) or (8.1.11) is given either in one of the sets E_1, E_2, E_3, E_4, E_5, E_6, or

a) in a certain moment of time, which we denote as $t = 0$, or

b) for $t \in \tau$, where τ is a sequence $t_1, t_2, \ldots$ converging to zero.

The number n, such that $N \leq n$ is given only in the cases $E = E_1, E_2, E_3$. In the case $E = E_1$ we suppose that the following condition (Ψ) holds:

(Ψ) each sum $\Phi_{j_1} + \Phi_{j_2} + \ldots + \Phi_{j_m} \neq 0$, where $j_1, j_2, \ldots, j_m$ is an arbitrary subset of the sequence $1, 2, \ldots, N$.

Besides, we suppose that each sequence P_{Π_i} which belongs to the set $E = E_1, E_2, E_3, E_4$ has the property A_μ defined in Subsection 8.10.2.

We need to find the numbers N, k_j, A_j, θ_j and the coordinates of the points M_j for each $j = 1, 2, \ldots, N$.

Suppose $B_\mu, O_{\Pi_1,\mu}, O_{\Pi_2,\mu}, \mu_{1,2}$ are the property and the sets defined in § 10.2.

Theorem 8.11.1. *In the case (a), the numbers N, A_j, θ_j and the points M_j ($j = 1, 2, \ldots, N$), and in the case (b) all the desired values are determined uniquely in the inverse problem 8.2 if we set the field u of form* (8.1.10) *or* (8.1.11) *in each of the sets E_3, E_4, E_5, E_6. If we set the field u in each of the sets E_1, E_2 then*

1) we can determine (in both cases a) and b)) whether a pair of straight lines Π_1, Π_2 satisfies the property B_μ. Here, in the case $E = E_2$ we denote by Π_1 and Π_2 two straight lines from the set π_2 but not from the set Π_μ. If the property B_μ holds, then, in the case a), we find uniquely the numbers

N, A_j, θ_j and the points M_j for each $j = 1, 2, \dots, N$. In the case b) we find all the unknown values. If the property B_μ fails, then, in both cases a) and b), we find the sets $O_{\Pi_1,\mu}$, $O_{\Pi_2,\mu}$ uniquely and a certain finite set $\mu_{1,2}$ consisted of N_1 points from the half-space $z > 0$ $(N \leq N_1 < \infty)$ (this set contains the desired set μ) and the collection of arbitrary sums

$$\Psi_{0j} = \Phi_{j_1} + \Phi_{j_2} + \ldots + \Phi_{j_m}.$$

Besides, in the case b) we can define the sums

$$\Psi_j(t) = \Phi_{j_1} e^{ik_{j_1} t} + \Phi_{j_2} e^{ik_{j_2} t} + \ldots + \Phi_{j_m} e^{ik_{j_m} t}$$

for $t \in \tau$, where $1 \leq m \leq N$, and $j_1, j_2, \dots, j_m$ are the numbers of those points M_j which lie in one semicircle from $O_{\Pi_1,\mu}$ or $O_{\Pi_2,\mu}$. Each point of the set $\mu_{1,2}$ lying in the semicircle from $O_{\Pi_1,\mu}$ or $O_{\Pi_2,\mu}$ which has no more points from $\mu_{1,2}$ is the point $M_j \in \mu$ (in both cases a) and b)). For this point M_j we determine uniquely, the numbers A_j, θ_j (in the case a)) and the numbers A_j, θ_j, k_j (in the case b)).

Besides, the set $\mu_{1,2}$ is such that in both cases a) and b), we can construct the third sequence of points P_{Π_3} from the plane $z = 0$ satisfying the following property. In both cases a) and b), the additional setting of the field u in P_{Π_3} for $t = 0$ allows us to find uniquely the numbers N, A_j, θ_j and the points M_j $(j = 1, 2, \dots, N)$; for $t = t_0 \neq 0$, — the numbers

$$N, A_j, \theta_{1j} = \arg(e^{ik_j t_0 + \theta_j}), \qquad \theta_{1j} \in [0, 2\pi)$$

(the initial phases in "new" time $t_1 = t - t_0$) and the points M_j $(j = 1, 2, \dots, N)$; for $t \in \tau$ we find all the desired values; and for $t \in \tau_0$, where τ_0 is the sequence converging to $t_0 \neq 0$, we find the numbers N, A_j, θ_{1j}, k_j and the points M_j $(j = 1, 2, \dots, N)$.

The proof of Theorem 8.11.1 and the method of solution of the inverse problem 8.2 are similar to those considerations of § 8.7 and Subsection 8.10.2. We should only replace k by k_j, $\Phi_j(k)$ by $f_j(t) = \Phi_j e^{-k_j t}$, $A_j(k)$ by Ψ_{0j} or by $\Psi_j(t)$ in formulas (8.10.1), (8.10.3) and (8.10.4). The number k_j in the case b) we define by the formula

$$\left.\frac{\mathrm{d}f_j}{\mathrm{d}t}\right|_{t=0} = ik_j \Phi_j.$$

Remark 8.11.1. We can obtain the other inverse problem from the inverse problem 8.1. For this purpose we change the field u of form (8.1.4) or (8.1.5) by reflected wave ψ of form (8.1.4) (for the half-space); we change the plane $z = 0$ by an arbitrary plane $z = H$ $(H > 0)$. All the sources are disposed in one half-space defined by this plane. The uniqueness theorem for the new inverse problem is obtained from Theorem 8.7.1 by the same change.

The analogous statement holds for the stationary problem also, where the role of reflected wave plays the second term in equality (8.1.10).

Remark 8.11.2. Suppose all the sources, i. e. the points M_j, lie in the half-space $z = 0$ or the one side from a certain straight line Π_0. Then the inverse problem 8.1 has a unique solution $\{N, M_j, m_j(j = 1, 2, \dots, N)\}$ in the static case and $\{N, M_j, \varphi_j(t), t_j(j = 1, 2, \dots, N)\}$ in the nonstationary case if instead of the set E we take the set consisted of $f(2n)$ arbitrary points in an arbitrary straight line parallel Π_0. The analogous statement holds for stationary problems also.

Remark 8.11.3. If in the inverse problem the coordinates of the points M_j are known, then, in order to find the values $\varphi_j(t)$ or m_j if suffices to take in a certain straight line from the plane $z = 0$ either $f(N)$ arbitrary points, or N points constructed by the set $\{M_j\}$.

8.12. POSSIBLE APPLICATIONS

The settings and the results stated above, evidently should be used in the situations when we are dealing with the set of vibration sources (mechanical, acoustic or electromagnetic), or point masses (the sources of gravitational field). In this case the situation is such that we can measure only the summary field generated by the set of sources. In other words, the field of each separate source and its coordinates and impulse characteristics we cannot measure in experience. However, we must determine these characteristics.

Such situation arises, first, in geophysics (in the inverse dynamic seismic problems) when we seek the sources of elastic waves inside the Earth.

These results may be an interest also, for example, in geophysical problem of vibrational radiating of the Earth and non-exploring seismic as the control of the work of many-element source of seismic waves (the vibrator, the impulse source) — the radiating seismic antenna. For example, we can use these results for estimation of synchronization of the impulses radiated by various vibrators (impulse sources) of one vibration block; for determination of phase shifts between the vibration blocks, which provide the directional radiation. We can use these results also for control of nonstability of the radiation parameter generated, in particular, by the change of the properties of the ground in the place of loading. We can use the results for determination of the phase shift between the constraining force and the vibration of the ground for each elementary source in the collection; for determination of mutual influence of the vibroelements (impulse sources) for their grouping, and the determination of the actual signals emanating from each element.

Second, such situation arises in acoustics when we determine the set of sound sources in the acoustic medium. For example, this is the problem studying of parameters and dynamics of the phenomena relating to the separate pulsing bubble under the ultrasonic cavitation (see Zhuravlyev and Akopyan, 1977). The analogous situation arises in the problem of the sound language of living organisms (for example, of social insects, in particular, bees, see Es'kov, 1977) in bioacoustics, where the problem of determining the sound impulse of each interacting organisms is of the great importance.

Third, such situation arises in certain problems of biophysics connected with biological radiations, where the sources of vibrations are the living cells. These are the problems connected with biological luminescence (Prosser and Brown, 1967; Chumakova and Gitel'son, 1975), superweak oscillation of biological systems (Tarusov, et al., 1967), the Gurvich oscillation (Gurvich, 1968), and the effect described in Kaznacheev, et al., (1974). Some bacteriums and simplest organisms radiate the light (Prosser and Brown, 1967; Chumakova and Gitel'son, 1975). Unlike the luminescence of living multicellular organisms, where the light impulses may play the role of signals (Prosser and Brown, 1967), it is not clear (Prosser and Brown, 1967; Chumakova and Gitel'son, 1975; Kaznacheev, et al., 1974) which informational meaning may have the radiation for living onecellular organisms — bacteria, protozoa and separate cells of multicellular organisms. In the experiments it is possible to measure both the radiation of separate cell (Chumakova and Gitel'son, 1975) and the summary radiation of the collection of the cells (for example, in culture). It is interesting to find, whether the radiation impulses is the communication language for cells. For this purpose it is important to know whether the signal of a separate cell differs from the signal of the cell in the whole population. It is difficult to answer this question directly since we can measure only the total signal.

To solve this problem, we may suggest the approach based on solving the inverse problems. In this case we can consider the inverse problems considered above as the simplest mathematical model of the inverse biological luminescence problems (in the wide scence), i. e. the determination of the coordinates (if they are unknown) and the characteristics of impulses of each separate emissive source given the summary oscillation field which is generated by the emissive source system (or by system of the oscillation nets, see Brajnes and Syslov, 1964). The obtained results show that it is possible in principle to find all these values. The basic difficulty here consists in the measurement of the field.

Radiation of living organisms has the following peculiarities:

1) The cells radiate the impulses which interrupt the periods of silence (Chumakova and Gitel'son, 1975; Kaznacheev, et al., 1974).

2) The number of cells may be rather large (for example, several millions or even milliards, Chumakova, and Gitel'son, 1975). Moreover, each cell may have several radiating particles (Prosser and Brown, 1967; Chumakova and Gitel'son, 1975).

3) The number of radiating cells may vary in time; for example, since their reproduction or dying.

Our mathematical model satisfies all these conditions. However, it is rather simplified relatively these phenomena and must be improved and approximated to the experiment conditions.

Bibliography

Alekseev, A. S. (1962). Inverse problems of wave propagation theory. I, II. *Izv. Akad. Nauk SSSR, Ser. Geophys.*, 11 (in Russian).

Alekseev, A. S. (1967). Inverse dynamical problems of seismology. In: *Some Methods and Algorithms for Interpretations of Geophysical Data.* Nauka, Moscow, 9–84 (in Russian).

Alekseev, A. S. and Megrabov, A. G. (1972). Direct and inverse problems of plane waves scattering from the inhomogeneous transition layers. In: *Mathematical Problems of Geophysics.* (3), Novosibirsk, 8–36 (in Russian).

Alekseev, A. S. and Megrabov, A. G. (1973). Direct and inverse problems of plane waves scattering from the inhomogeneous transition layers (hyperbolic and elliptic cases). In: *Theory of Diffraction and Wave Propagation.* Trudy VI Vsesojuznogo Simpoziuma po Difrakcii i rasprostraneniu voln. 2, Moscow-Erevan, 352–356 (in Russian).

Alekseev, A. S. and Megrabov, A. G. (1974). Inverse problems for a string with the directional derivative condition given in one end and the inverse problems of scattering the plane waves from the inhomogeneous layer. *Dokl. Alad. Nauk SSSR,* 219 (2), 308–310 (in Russian).

Alekseev, A. S. and Megrabov, A. G. (1977). Determination of the trajectory of impulse motion in the inverse problem for the wave equation. *Journ. of Calc. Math. and Math. Phys.* 17 (6), 1508–1522 (in Russian).

Ambarzumjan, W. A. (1929). Uber eine frage der Eigenwerttheorie. *Zeitschrift fur Fhysik,* 53, 690–695.

Anikonov, Yu. E. (1978). *Some Methods of Research of Multidimensional Inverse Problems for Differential Equations.* Nauka, Sib. Otdel., (in Russian).

Anikonov, Yu. E. (1995). *Multidimensional Inverse and Ill-Posed Problems for Differential Equations.* VNU Science Press, Utrecht, The Netherlands.

Antonenko, O. F. (1967). Finite-difference scheme inversion for solving a one-dimensional inverse problem in seismology. In: *Some Methods and Algorithms for Interpretation of Geophysical Data.* Computer Center of Siberian Branch of USSR. Academy of Sciences, Novosibirsk, 18–25 (in Russian).

Bargman, V. (1949). On the connection between phase shift and scattering potential. *Rev. Mod. Phys.* 21, 488.

Belishev, M. I. (1975). On violation of solvability condition for an inverse problem for inhomogeneous string. In: *Functional Analysis and its Application,* 9 (4), 57–58 (in Russian).

Belishev, M. I. (1977). The phenomena of complete internal reflection in the problem of incidence of the inclined wave. In: *Nonclassical Methods of Geophysics.* Computer Center of Siberian Branch of USSR, Academy of Sciences, Novosibirsk, 130–131 (in Russian).

Belishev, M. I. (1978). *Some Singular Inverse Problems of the Theory of Wave Propagation.* Autoreferat of dissertation, Leningrad (in Russian).

Belishev, M. I. (1979). Inverse problem for scattering the plane waves for one class of layered media. *Zapiski Nauchn. Seminarov LOMI*, 78, 30–53 (in Russian).

Belishev, M. I. and Blagoveshchenskii, A. S. (1999). *Dynamical Inverse Problems of the Wave Theory.* Izd. Sankt-Peterburg Gos. University (in Russian).

Bitsadze, A. V. (1966). *Boundary-Value Problems for Second Order Elliptic Equations.* Nauka, Moscow (in Russian).

Blagoveshchenskii, A. S. (1966). On the inverse problem in the theory of propagation of seismic waves. In: *Problems of Mathematical Physics,* 1, Leningrad University Press, Leningrad, 68–81 (in Russian).

Blagoveshchenskii, A. S. (1970a). An inverse problem for the wave equation with an unknown source. In: *Problems of Mathematical Physics,* 4, Leningrad University Press, Leningrad, 27–39 (in Russian).

Blagoveshchenskii, A. S. (1970b). On various setting of a one-dimensional inverse problem for the telegraph equation. In: *Problems of Mathematical Physics,* 4, Leningrad University Press, Leningrad, 40–41 (in Russian).

Blagoveshchenskii, A. S. (1971). The local method of solution of the non-stationary inverse problem for an inhomogeneous string. *Proceedings of the Mathematical Steklov Institute,* 115, 28–38 (in Russian).

Blagoveshchenskii, A. S. and Buzdin, A. A. (1972). A setting of the boundary-value inverse problem for the string equation in the case when the properties of the string are described by the rapidly vibrating functions. *Trudy Vsesojuzn. Simp. po Obratn. Zadacham,* Novosibirsk, 31–35 (in Russian).

Blagoveshchenskii, A. S. and Voevodskii, K. E. (1981). Inverse problem of scattering from the layered-inhomogeneous half-space. *Differential Equations XVII,* (8), 1434–1445 (in Russian).

Bleistein, N. and Cohen, J. K. (1977). *Journ. Math. Phys.* 18 (2), p. 194.

Borg, G. (1945). Eine Umkehrung der Sturm–Liouvillschen Eigenwertaufgabe. Bestimmung der Differentialgleichund durch dir Eigenwerte, *Acta Math.,* 78 (2,1).

Borodaeva, N. M. (1969). On a numerical solution of one-dimensional dynamical problem in the scheme of exploration of marine sediments. *Mathematical Problems of Geophysics,* Vyp. I, Computer Center of Siberian Branch of USSR, Academy of Sciences, Novosibirsk, 225–234 (in Russian).

Brajnes, S. N. and Suslov, A. I. (1964). Informational processes in biocybernetics. Experimental Surgery and Anaesthesiology, (2), 13–18 (in Russian).

Brekhovskih, L. M. (1973). *Waves in Layered Media.* Nauka, Moscow (In Russian).

Bukhgeim, A. L. (1999). *Volterra Equations and Inverse Problems.* VNU Science Press, Utrecht, The Netherlands.

Bukhgeim, A. L. (2000). *Introduction to the Theory of Inverse Problems.* VNU Science Press, Utrecht, The Netherlands.

Bukhgeim, A. L. Zerkal, S. M., Konev, V. T., and Sabitora, G. S. (1985). On a class of inverse problems in the discrete setting. In: *Inverse Problems of Mathematical Physics,* Computer Center of Siberian Branch of USSR, Academy of Sciences, Novosibirsk, 57–65 (in Russian).

Bukhgeim, A. L., Konev, V. T., and Sabitova, G. S. (1995). *Uniqueness and Stability of Inverse Problems of Solvability of Signals in the Discrete Setting.* Preprint (5). Institute of Mathematics of Siberian Academy of Sciences. Novosibirsk (in Russian).

Chudov, L. A. (1949). Inverse Sturm–Liouville problem. *Mathemat. Sbornik,* 25 (3), 451–456 (in Russian).

Chumakova, R. I. and Gitel'son, I. I. (1975). *Luminous Bacteria.* Nauka, Moscow (in Russian).

Dobrinskii, V. I. and Gorbunov, V. A. (1977). Some problems of scattering of plane SH-waves on one-dimensional inhomogeneous media. In: *Calculating Problems of Mathematical Problems of Geophysics.* Computer Center of Siberian Branch of USSR, Academy of Sciences, Novosibirsk, 48–63 (in Russian).

Es'kov, E. K. (1977). *Successes of Modern Biology.* 83 (3), (in Russian).

Faddeev, L. D. (1959). Inverse problem of quantum theory of scattering. *Uspekhi Math. Nauk,* 19 (4), 57–119 (in Russian).

Fikhtengolts, G. M. (1968). *Basis of Mathematical Analysis.* Nauka, Moscow (in Russian).

Gasymov, I. M. and Levitan, B. M. (1964). Determination of differential operator from its two spectra. *Uspekhi Math. Nauk,* 19 (2), 3–63 (in Russian).

Gel'fand, I. M. and Levitan, B. M. (1951). On the determination of a differential equation from its spectral function.*Izv. Acad. Nauk SSSR, Ser. Mathem.* 15, 309–360 (in Russian).

Gerver, M. L. (1970). Inverse problem for the string equation. *Izv. Acad. Nauk SSSR, ser. Fizika Zemli,* 8, 3–20 (in Russian).

Girand, G. (1932). Generalization des problemes sur les operations du type elliptique, *Bull. sc. math.,* 56, 248–272, 281–312, 316–352.

Gurvich, A. A. (1968). *Problem of Mitogenetic Radiation as the Aspect of Molecular Biology.* Medicine, Leninngrad (in Russian).

Hopf, E. (1927). Elementare Betrachtungen uber die Losungen partiellen Differentialgleichtungen zweiter Ordnung vom elliptischen Typus. *Sitzungb. Preuss. Acad. Wiss.,* 19, 147–152.

Hopf, E. (1952). A remark on linear differential equations of second order. *Proc. Amer. Math. Soc.* 3, 791–793.

Isakov, V. M. (1977). *Dokl. Akad. Nauk SSSR,* 237 (2), p. 257 (in Russian).

Iskenderov, A. D. (1972). An inverse problem for equations of elliptic type. In: *Inverse Problems for Differential Equations.* Computer Center of Siberian Branch of USSR, Academy of Sciences, Novosibirsk, 107–115 (in Russian).

Jost, R. (1947). Uber die balschen Nullstellen der Eigenwerte der S-Matrix, *Helv. Phys. Asta,* 20, 256.

Kabanikhin, S. I. (1979). On solvability of the inverse dynamical seismic problem. In: *Conditionally-Correct Mathematical Problems and Problems of Geophysics.* Computer Center of Siberian Branch of USSR, Academy of Sciences, Novosibirsk, 43–51 (in Russian).

Kabanikhin, S. I. (1984). On solvability of inverse problems for differential equations. *Dokl. Acad. Nauk SSSR,* **277** (4), 788–791 (in Russian).

Kabanikhin, S. I. (1988). *Projection-Difference Methods for Determining the Coefficients of Hyperbolic Equations.* Nauka, Siberian Branch, Novosibirsk (in Russian).

Kabanikhin, S. I. and Lorenzi, A. (1999). *Identification Problems of Wave Phenomena. Theory and Numerics.* VNU Science Press, Utrecht, The Netherlands.

Kardakov, V. B. (1976). *Ill-Posed Mathematical Problems and Geophysical Problems.* Izd. Vych. Tsentr Sib. Otdel. Akad. Nauk SSSR, Novosibirsk, 46–63 (in Russian).

Karlin, S. and Studden, W. I. (1973). *Tchebysheff Systems: with Applications in Analysis and Statistics.* Interscience Publishers, A. Division of John Wiley and Sons, New-York-Leningrad-Sydney.

Kats, I. S. and Krein, M. G. (1968). On spectral functions of the string. In: Atkinson, F. V. (1964). *Discrete and Continuous Boundary Problems.* Academic Press, New-York—London. Also translated "Mir", Moscow, 1968 (in Russian). Complement 2. 648–737.

Kaznacheev, V. P., Mikhailova, L. P., and Shurin, S. P. (1974). Information interactions in biological systems stimulated by electromagnetic radiation of optical range. In: *Progress of Biological and Medicine Cybernetics.* Medicine, Moscow, 314–338 (in Russian).

Koshlyakov, N. S., Gliner, E. B., and Smirnov, M. M. (1970). *Equations in Partial Derivatives of Mathematical Physics.* Vysh Shkola, Moscow (in Russian).

Krein, M. G. (1951a). Solution of the inverse Sturm-Liouville problem. *Dokl. Akad. Nauk SSSR,* 76 (1), 21–24 (in Russian).

Krein, M. G. (1951b). Determination of density of inhomogeneous symmetric string from the spectra of its frequencies. *Dokl. Akad. Nauk SSSR,* 76 (3), 345–348 (in Russian).

Krein, M. G. (1952). On inverse problems for inhomogeneous string. *Dokl. Akad. Nauk SSSR,* 82 (5), 669–672 (in Russian).

Krein, M. G. (1953). On transition function of a one-dimensional boundary-value problem of the second order. *Dokl. Akad. Nauk SSSR,* 88 (3), 405–408 (in Russian).

Krein, M. G. (1954). On a method of effective solution of an inverse boundary problem. *Dokl. Akad. Nauk SSSR*, 94 (6), 767–770.

Langer, R. E. (1931). On the asymptotic solutions of ordinary differential equations with an application to the Bessel functions of large order. *Transactions of the American Math. Society,* 33 (1), 23–64.

Lavrentiev, M. M. (1972). Inverse problems for differential equations. In: *Inverse Problems for Differential Equations.* Nauka, Siberian Branch, Novosibirsk, 7–13 (in Russian).

Lavrentiev, M. M., Romanov, V. G., and Vasil'ev, V. G. (1967). *Multidimensional Inverse Problems.* Lecture Notes of Mathematics, 167, Springer-Verlag, New-York.

Lavrentiev, M. M., Romanov, V. G., and Shishatskii, S. P. (1986). *Ill-Posed Problems of Mathematical Physics and Analysis.* American Mathematical Society, Providence, Rhode Island.

Lavrentiev, M. M. and Shabat, B. V. (1973). *Methods of Theory of Functions of Complex Variable.* Nauka, Moscow (in Russian).

Leibenzon, L. S. (1942). *Brief Course of Elastic Theory,* Moscow-Leningrad, Gostehizdat (in Russian).

Levinson, N. (1949). On the uniqueness of the potential in a Schroedinger equation for a given asymptotic phase. *Kgl. Donske Videnskab Selskab mat.-fis. medd.* 25 (9).

Levitan, B. M. (1962). *Operators of Generalized Shift and its Applications.* Fizmatgiz, Moscow (in Russian).

Levitan, B. M. and Sargsyan, I. S. (1970). *Introduction in the Spectral Theory,* Nauka, Moscow (in Russian).

Marchenko, V. A. (1950). Some problems of theory of differential operator of the second order. *Dokl. Akad. Nauk SSSR,* 72 (3), 457–460 (in Russian).

Marchenko, V. A. (1952). Some problems of theory of one-dimensional linear differential operators of the second order. *Trudy Mosc. Math. Obshch.* 1, 327–420 (in Russian).

Megrabov, A. G. (1972). Direct and inverse problems of scattering of plane waves on inhomogeneous transition layers (elliptic case). In: *Mathematical Problems of Geophysics.* Vyp. 3. Computer Center of Siberian Branch of USSR, Academy of Sciences, Novosibirsk, 113–123 (in Russian).

Megrabov, A. G. (1973a). Inverse problems of scattering the plane waves on the inhomogeneous layers with free or fixed boundaries (hyperbolic case). In: *Mathematical Problems of Geophysics.* Vyp. 4. Computer Center of Siberian Branch of USSR, Academy of Sciences, Novosibirsk, 84–102 (in Russian).

Megrabov, A. G. (1973b). Inverse problems for elliptic equation in the strip connected with the problem of scattering the plane waves on the inhomogeneous layers. In: Mathematical Problems of Geophysics. Vyp. 4, Computer Center of Siberian Branch of USSR, Academy of Sciences, Novosibirsk, 103–115 (in Russian).

Megrabov, A. G. (1973c). Inverse problems for hyperbolic and elliptic equations with the data in a countable set of solutions and inverse problems of scattering the plane waves. In: *Mathematical Problems of Geophysics,* Vyp. 4. Computer Center of Siberian Branch of USSR, Academy of Sciences, Novosibirsk, 116–130 (in Russian).

Megrabov, A. G. (1974). The way of recovery of density and velocity in inhomogeneous layer as the function of the depth by giving the set of plane waves reflected from the layer under various angles. In: *Mathematical problems of geophysics,* Vyp. 5, p.II. Computer Center of Siberian Branch of USSR, Academy of Sciences, Novosibirsk, 78–107 (in Russian).

Megrabov, A. G. (1975a). Inverse problems for elliptic equations in half-plane and in the strip and inverse problems of scattering the plane waves in inhomogeneous layers. *Dokl. Acad. Nauk SSSR,* 220 (2), 315–317 (in Russian).

Megrabov, A. G. (1975b). Inverse problems for the equation of mixed type. In: *Mathematical Problems of Geophysics.* Vyp. 6, pt I, Computer Center of Siberian Branch of USSR, Academy of Sciences, Novosibirsk, 122-143 (in Russian).

Megrabov, A. G. (1977a). Some inverse problems for an equation of mixed type. *Soviet Math. Dokl.,* 18 (3), 662–665.

Megrabov, A. G. (1977b). Inverse problems of determining an arbitrary set of point sources. *Dokl. Akad. Nauk SSSR,* 237 (5), 1063–1065 (in Russian).

Megrabov, A. G. (1978a). Inverse problems for equation of mixed type. In: *Methods of Functional Analysis in Problems of Mathematical Physics.* Izd. of Inst. of Math. Akad. Nauk Ukr. SSR, Kiev, 23–29 (in Russian).

Megrabov, A. G. (1978b). On determination of the set of point sources. *Dokl. Akad. Nauk SSSR,* 242 (6), 1303–1306 (in Russian).

Mikhailova, N. G. and Parijskii, B. S. (1969). An inverse problem for the wave equation with a surface source. In: *Calculating Seismology.* Vyp. 4, 95–138 (in Russian).

Miranda, C. (1957). *Equations with Partial Derivatives of Elliptic Type.* Moscow (in Russian).

Naymark, M. A. (1954). Spectrum research and expansion by eigenfunctions of not self-adjoint differential operator of the second order on the half-axis. *Trudy Mosc. Math. Obshch.* 3, 181–270 (in Russian).

Naymark, M. A. (1956). Spectral analysis of not self-adjoint operators. *Uspekhi Math. Nauk SSSR,* 11 (6), 183–202 (in Russian).

Naymark, M. A. (1969). *Linear Differential Operators.* Nauka, Moscow (in Russian).

Nikol'skii, E. V. (1964). Reflection of plane elastic waves from an inhomogeneous layer in the case of normal falling. *Appl. Math. and Techn. Phys.* 4, 65–74 (in Russian).

Nikol'skii, E. V. (1965). Reflection of plane nonstationary waves from an arbitrary inhomogeneous half-space. Acoustic case. *Appl. Math. and Techn. Phys.* 3, 63–67 (in Russian).

Oleinik, O. A. (1952). Properties of solutions of some boundary-value problems for equations of elliptic type. *Math. Sbornik,* 30, 695–702 (in Russian).

Parijskii, B. S. (1969). Inverse problem for the wave equation with the action in the depth. In: *Calculating Seismology.* Vyp. 4, 139–169 (in Russian).

Petrovskii, I. G. (1951). *Lectures in Theory of Integral Equations.* Moscow-Leningrad, Gostechizdat (in Russian).

Polya, G. (1922). On the mean value theorem corresponding to a given linear homogeneous differential equation. *Trans. Am. Math. Soc.,* 24, 312–324.

Prilepko, A. I. (1973). Inverse problems of potential theory. *Math. Zametki,* 14 (5), 755–765 (in Russian).

Prosser, L. and Broun, F. (1967). *Comparative Physiology of Animals.* Mir, Moscow (in Russian).

Remez, E. Ya. (1969). *The Basis of Numerical Methods of Chebyshev Approximation.* Kiev, Naukova Dumka (in Russian).

Romanov, V. G. (1972a). The problem of determining of one-dimensional velocity of signal propagation in the half-space from the regime of vibrations of one point of this space. In: *Mathematical Problems of Geophysics,* Vyp. 3. Computer Center of Siberian Branch of USSR, Academy of Sciences, Novosibirsk, 164–186 (in Russian).

Romanov, V. G. (1972b). *Some Inverse Problems for Equations of Hyperbolic-Type,* Nauka, Siberian Branch, Novosibirsk (in Russian).

Romanov, V. G. (1973). *Inverse Problems of Differential Equations.* Izd. Novosib. Gos. Univ., Novosibirsk (in Russian).

Romanov, V. G. (1987). *Inverse Problems of Mathematical Physics.* VNU, Science Press, Utrecht, The Netherlands.

Romanov, V. G. and Kabanikhin, S. I. (1991). *Inverse Problems of Geoelectrics.* Nauka, Moscow (in Russian).

Romanov, V. G. and Kabanikhin, S. I. (1994). *Inverse Problems for Maxwell's Equations.* VNU, Science Press, Utrecht, The Netherlands.

Shilov, G. E. (1970). *Mathematical Analysis.* Part 3, Nauka, Moscow (in Russian).

Smirnov, M. M. (1970). *Equations of Mixed Type.* Nauka, Moscow (in Russian).

Sylvester, J. and Uhlmann, G. (1987). A global uniqueness theorem for the inverse boundary-value problem. *Annals of Mathematics,* 125, 153–169.

Tarusov, B. N., Ivanov, I. I., and Petrusevich, Yu. M. (1967). *Superweak Luminescence of Biological Systems,* Izd. Mosc. God. Univ., Moscow (in Russian).

Tikhonov, A. N. and Samarskii, A. A. (1966). *Equations of Mathematical Physics,* 3 ed., Nauka, Moscow (in Russian).

Titchmarsh, E. K. (1948). *Introduction in the Theory of Fourier Integrals.* Leningrad-Moscow, Gostehizdat (in Russian).

Zaremba, S. (1927). Sur un probleme toujours possible comprenant, a titre de cas particuliers, le probleme de Dirichlet et celui de Neumann. *J. Math. Pures Appl.*, 6, 127–163.

Zapreev, A. S. (1976). Uniqueness theorem for solution of one plane inverse problem for the Helmholtz equation. *Ill-Posed Mathematical Problems and Geophysical Problems,* Novosibirsk, 46–63 (in Russian).

Zapreev, A. S. and Tsetsokho, V. A. (1976). *Inverse Problems for the Helmholts Equation.* Preprint 22. Computer Center of Siberian Branch of USSR. Academy of Sciences. Novosibirsk (in Russian).

Zhuravl'ev, A. I. and Akopyan, V. B. (1977). *Ultrasonic Luminescense.* Nauka, Moscow (in Russian).